PRAISE FOR *CLEAN WATER USING SOLAR AND WIND: OUTSIDE THE POWER GRID*

"Prof. Olsson has deep insights and wisdom on the importance of synergy between water and energy sectors, thus this book would serve as a main reference for professionals on how both sectors could serve the needs of the people in a sustainable manner!"

Dato' Seri Ir. Dr. Zaini Ujang
Secretary-general, Energy, Green Technology and Water Ministry (KeTTHA), Malaysia

"Too many people still lack access to clean water and energy. Professor Gustaf Olsson skilfully explains how modern clean energy options could bring clean water to less privileged people all over the world. This book truly deepens our understanding of the water-energy nexus."

Professor Peter D. Lund
Aalto University, Finland

"Prof. Olsson's latest book is designed to enhance understanding of the renewable energy revolution that is now underway, and how use of solar and wind energy can help provide water services to those currently underserved. He succeeds in a carefully written, concise, and yet comprehensive book that will appeal to many audiences. These include water professionals who wish to learn about solar and wind energy, public and private officials responsible for providing energy and water services, those in the financial community who will be called upon to provide the necessary investment funds, and those students interested in entering these emerging fields. His book is an important contribution to understanding vital changes taking place broadly in the 21st century that will improve the lives of millions."

Dr. Allan Hoffman
Senior Analyst, Office of Energy Efficiency and Renewable Energy, U.S. Department of Energy (retired), Author, *The U.S. Government and Renewable Energy: A Winding Road*, Washington D.C., United States

"Water and energy are two of the most important pillars that support modern civilisation. Despite many similarities, the two have traditionally been generated and delivered in very different ways by different organisations and subject to different legislation and regulation but to the same customers. The very success of conventional, centralised distribution systems in large urban areas has also been their limitation for communities not living in industrialised countries. Cost and scale. This book explores and supports the provision of clean water to communities which are off-grid based on the tremendous advances made in recent years around renewable energy generation, especially solar PV and wind. It explains in the clearest possible terms the potential, progress, economics and technology that can make this a reality. It is a book for specialists and a book for generalists written in a clear and engaging style by an author of immense experience, knowledge and wisdom. It can't be said for many books, but this really is one that can change the world!"

Professor David Butler
Director, Centre for Water Systems, University of Exeter, UK

"Gustaf is certainly a leader in the area of the water-energy nexus, having previously published one of the most useful books on this topic – *Water and Energy: Threats and Opportunities*. In his new book, Gustaf makes a clear case for the value of off-grid water and sanitation systems and the enormous potential for renewables to play a part in these systems. Gustaf has a special skill when it comes to explaining concepts in an understandable way and clearly shows how solar and wind can be integrated at each stage of the water supply and sanitation process."

Professor Shuming Liu
Vice Dean of School of Environment, Tsinghua University,
Beijing, China

"Following in-depth discussions on the water and energy nexus in his previous works (Water and Energy, Olsson 2012, 2015, IWA Publishing), Gustaf shows, in this new book, how renewable energy in combination with decentralized water operations can decouple many of these dependencies and meet challenges caused by climate change, population increase, water scarcity and poor water quality. Gustaf has rightfully stressed the specific importance of solar and wind energy for providing clean drinking water and sanitation to the underprivileged communities. This book will be a highly valuable reference for both power and water engineers.

With a strong background in both power and water engineering, Gustaf is uniquely positioned to address the water and energy nexus. I would like to thank, and congratulate, Gustaf for his continued thought leadership, even 12 years after his retirement!"

Professor Zhiguo Yuan
Director, Advanced Water Management Centre, the University of
Queensland, Brisbane, Australia

"This is a timely and necessary book as countries are starting to address the SDGs. Water and energy are crucial to development and decentralized or off-grid water and energy systems are vital to the developing world. Most of rural Africa will remain unserved unless these systems are implemented, due to economic constraints in bringing conventional grid-based services over long distances. I particularly enjoyed Chapter 2 showing the coupling between the different SDGs, it addresses my fear that we tackle the SDGs in isolation which was not the intent and this chapter brings the complexity home. Lastly, the book is an easy read for both the lay person and the scientist, a rare feat in itself."

Henry J. Roman
Director, Environmental Services and Technologies
Department of Science and Technology, Pretoria, South Africa

Clean Water Using Solar and Wind

Clean Water Using Solar and Wind

Outside the Power Grid

Gustaf Olsson

Published by

IWA Publishing
Alliance House
12 Caxton Street
London SW1H 0QS, UK
Telephone: +44 (0)20 7654 5500
Fax: +44 (0)20 7654 5555
Email: publications@iwap.co.uk
Web: www.iwapublishing.com

First published 2018

British Library Cataloguing in Publication Data
A CIP catalogue record for this book is available from the British Library

ISBN: 9781780409436 (Paperback)
ISBN: 9781780409443 (eBook)

Contents

Chapter 2
Towards sustainability goals **19**

Chapter 3
The renewable energy revolution   **27**

Part II
Water Technologies **45**

Chapter 4
Water supply **47**

Chapter 13
Land use for energy *159*

Chapter 14
Water operations using renewables – some cases *163*

Part V
The Future *173*

Chapter 15
Outlook to 2030 and further *175*

Acronyms

AC	alternating current
AD	anaerobic digestion
AMTA	American Membrane Technology Association
BOD	biochemical oxygen demand, a measure of the organic carbon content in wastewater
BWRO	brackish water reverse osmosis
Capex	capital expenditures
COD	chemical oxygen demand
CSP	concentrating solar power
DC	direct current
DO	dissolved oxygen
DoD	depth of discharge
EPBT	energy payback time
GHI	global horizontal irradiance
GOGLA	Global Off-Grid Lighting Association
GWP	global warming potential
IAEA	International Atomic Energy Agency
IEA	International Energy Agency
IEC	International Electrotechnical Commission
IRENA	International Renewable Energy Agency

IWR	irrigation water requirement
LCOE	levelised cost of energy
LCOS	levelised cost of storage
LED	light-emitting diode
LOHC	liquid organic hydrogen carrier
MCI	manufacturing, construction and installations
MED	multiple effect distillation
MF	microfiltration
MSF	multistage flash distillation
NF	nanofiltration
NOM	natural organic material
NRECA	National Rural Electric Cooperative Association
NREL	National Renewable Energy Laboratory
O&M	operations and maintenance
Opex	operating expenditures
PAYG	pay as you go
ppm	parts per million or mg/l
PV	photovoltaic
REEEP	Renewable Energy and Energy Efficiency Partnership
RO	reverse osmosis
SDG	(The United Nations) Sustainable Development Goals
SHS	solar home system
SSA	sub-Saharan Africa
SSD	solar still distillation
STC	standard testing conditions
SWH	solar water heating
SWRO	seawater reverse osmosis
SWT	small wind turbines
TRL	technology readiness level
UF	ultra-filtration
UNDP	United Nations Development Programme
UNESCO	United Nations Educational, Scientific and Cultural Organization

UNICEF	United Nations Children's Fund, initially UN International Children Emergency Fund
UNIDO	United Nations Industrial Development Organization
USAID	United States Agency for International Development
VFB	vanadium redox-flow batteries
WEC	World Energy Council
WEF	World Economic Forum
WHO	World Health Organization
WWDR	(United Nations) World Water Development Report

Preface and Guide for the Reader

With this book I hope to raise awareness about the technology revolution that can make both water and energy obtainable for all. Still more than a billion (=10^9; note that billion in Europe means 10^{12}) people lack access to both clean water and clean energy. Today renewable energy is becoming affordable for the underprivileged. Solar and wind energy are abundant in many regions outside the national power grids. Decentralised water treatment technology is available.

The book is aimed at different categories of readers:

- The **water professional** who wishes to learn about renewable energy: you can skip some of the water technology descriptions;
- The **power engineer** looking for applications of renewable off-grid energy: you can omit the basic information about solar PV and wind turbines;
- The **policy-maker**: no need to understand all the technical details. Still you can appreciate the enormous potential of off-grid renewable energy for water operations;
- The **investor**: this is one of the most meaningful investments that you can make;
- The **student**: the future energy source is renewable. This has huge consequences for water supply. You must be familiar with it.

Energy has always fascinated me and may explain why I wanted to study nuclear engineering in the 1960s, a time when peaceful nuclear energy was supposed to save the world. I was engaged in the planning of the first Swedish nuclear reactor. Soon I became involved in automatic control and got a faculty position at the Department of Automatic Control at Lund Institute of Technology (now the Engineering Faculty of Lund University), Sweden in 1967. As a control engineer I was challenged in 1973 to discover whether control could be of any value for wastewater treatment operations. This triggered my interest in water and over the years I have been increasingly involved in water system operational challenges. In the Department of Industrial Automation at Lund University we did research applying control and automation in water, power and electric energy systems.

When I retired in 2006 and had more time for reflection, I started to see more clearly the many connections between water and energy and how closely they depend on each other. The buzzword water-energy nexus had been created. Dr Allan Hoffman, at that time Senior Analyst at the US Department of Energy, Washington D.C., was probably the first to use the term. He had opened my eyes to the water-energy challenges and we met in person for the first time in Washington D.C. in 2008. Since then we have had regular contact, and Allan has given me a lot of constructive feedback, new insights and encouragement.

At the time when the first edition of my book *Water and Energy* was published in 2012 the challenges of the water-energy nexus had been widely recognised. I was quite pessimistic about the development of the climate negotiations, water quantity and quality consequences of fossil fuel exploration and processing, oil accidents and oil spills, and the lack of political will to make any positive changes towards a more sustainable future. However, seeing progress in the climate negotiations invigorated my spirits and provided inspiration for the second edition of the book (Olsson, 2015). The commitments by both the US and China to sign the Paris Agreement were a truly positive sign.

In the last chapter of Olsson (2015) I tried to describe the new hope from renewable energy and the possibility that water and energy can be decoupled for energy production. So, in 2017 I was encouraged by Mark Hammond, IWA Publishing, to widen the scope of the chapter and examine how renewable energy can provide water, not only in areas where electricity is already available but also in remote regions in developing parts of the world outside existing electric power grids.

Solar photovoltaic (PV) and wind energy have an enormous potential to bring electricity anywhere and to improve quality of life for millions of people. Solar PV has already exceeded all expectations. As a result, international agencies and organisations like the World Bank, IEA (International Energy Agency) and IRENA (The International Renewable Energy Agency) are continuously raising their predictions of future growth. China is leading the world in solar PV and wind turbine installations. The German Energiewende has triggered a remarkable development of renewable energy. Solar home systems have already been installed in millions of homes. All this progress has led to huge development in manufacturing skill and quality, which will increase the potential for use in other parts of the world.

There is an encouraging development in the efforts to increase the speed of off-grid electrification. In 2012 the Global Off-Grid Lighting Association (GOGLA, 2017) was established, an independent, not-for-profit industry association (www.gogla.org). GOGLA represents over 100 members as the voice of the off-grid solar energy industry. The organisation was born out of the IFC/World Bank's Lighting Global programme. Lighting Global (www.lightingglobal.org) is the World Bank Group's platform to support sustainable growth of the international off-grid solar market.

In the 1960s the activists chanted "power to the people". They never dreamed of the innovations that could spread light to the darkest corners of Earth. We now witness an energy revolution that has the potential to change quality of life for the world's most disadvantaged and poor. Distributed energy will change the relationship between the producer and consumer and will empower the powerless.

To supply renewable energy outside the existing electric power grids is the key not only to achieving universal access to electric energy, but also to bringing water to the millions of people who have no access to clean water today.

It is my ambition and hope that the book will raise the appreciation of the new possibilities that renewable energy gives in providing clean water for all. It is the combination of pressure from the grassroots movements and concerned citizens in combination with technology development, decreasing costs and interest from policy and decision makers that will make this dream a reality.

Gothenburg, Sweden
May 2018
G.O.

Acknowledgements

It is quite a privilege and gift having several friends and colleagues as support. Professional colleagues have helped me to be honest and friends have supported me to keep hold of the inspiration to write the book.

Ever since I met Allan Hoffman in person at the Department of Energy in Washington D.C. in 2008 he has generously shared with me both his knowledge and his encouragement. He has suggested several changes and additions to the manuscript. Gianguido Piani, my co-author on the book *Computer Systems for Automation and Control*, translated into both German and Russian, has given me a lot of constructive feedback. Zhiguo Yuan, Director, Advanced Water Management Centre, The University of Queensland, Brisbane, has offered several good suggestions. Shuming Liu, Professor at Tsinghua University, Beijing, has verified important facts. Lawrence Jones, Vice President, International Programs, Edison Electric Institute, Washington D.C., has recognised the crucial connections between water and energy. Having grown up in Liberia, he not only knows about African conditions but is engaged in many activities to promote technology transfer to African regions. He has pointed out many critical aspects of water-energy issues for Africa. Jörgen Svensson, Lund University, made sure that I have described wind power correctly.

Mike Greenberg helped me to improve my language. Ove Finndin, highly engaged in environmental issues, encouraged me in my effort.

Mark Hammond, Books Commissioning Editor at IWA Publishing, inspired me to extend and update the last chapter of my previous book *Water and Energy* from 2015 into a new publication. There has been a hugely significant development of renewable energy only in the last three years, and this will influence the access to clean water. The staff at IWA Publishing are always so positive. Thank you, Niall Cunniffe, Zoë Dann and Jacqui Lewis, for your friendly support and professional help.

Only a week before the manuscript was submitted to the publisher the content was tested at a three-day workshop with 16 PhD students in Sweden. They gave me encouragement that the text is timely and relevant.

Kirsti, my wonderful wife: I am often astonished that you can endure my odd habits of work. You not only love me and give me never-ending encouragement; you also give me meaningful questions, valuable feedback and help me to hold on to a more complete perspective of life.

Part I

Water and Energy – A Human Right

Clean water and basic energy are needed for a decent life. This is taken for granted by most people in the world. Yet still, in 2018, more than 650 million people lack clean water and around 1,000 million people do not have the electric power that could enable them to light their homes, cook their food and access clean water.

This book addresses this issue and claims that there are better opportunities than ever to satisfy the basic needs of clean water and energy for all.

The unique development of renewable energy over the last few years brings new hope for millions of people. Those living in remote areas, outside the national electric power grids, will now have a realistic opportunity to receive the benefits of clean energy. This in turn will open up possibilities of obtaining clean water and enable people to escape poverty, fight hunger, improve health and support education.

The United Nations has defined 17 Sustainable Development Goals (SDGs). Two of these relate to clean water and clean energy. However, without these two essentials it will hardly be possible to reach the other 15 goals.

Chapter 1

Water and energy – for all

The aim of this book is to describe how solar photovoltaic (PV) and wind energy have a huge potential to supply clean water for the developing world, in particular in areas with no electric power grid connection. Off-grid technologies can form a significant part of the solution, all the way from household level to village or community level. Small-scale off-grid systems can provide not only lighting but also energy for pumping to gain access to water and for treatment to purify and reuse water.

The cost development of renewable energy has been remarkable and will make electric power affordable even for the poorest. Since 2010, the cost of key energy devices has declined dramatically: LED lighting is 95% cheaper, solar PV 60% and battery storage 75% less expensive.

© IWA Publishing 2018. Clean Water Using Solar and Wind: Outside the Power Grid
Author: Gustaf Olsson
doi: 10.2166/9781780409443_0003

Already today the cost of "new" renewable energy can compete with traditional electrical generation.

Clean water is a matter of life and death. Still, too many people lack this basic need. Lack of electric power for pumping and cleaning contaminated water is one of the missing prerequisites; one person out of seven has no electric power available. We wish to raise awareness of the fact that today there are great and realistic opportunities for those people living outside the electric power grid.

1.1 CLEAN WATER AND ENERGY FOR ALL

The World Economic Forum (WEF) presented its tenth global risk report in 2015 (http://reports.weforum.org/global-risks-2015). For the first time water crises took the top spot among the risks in the report, published by 900 leaders in politics, business and civic life about the world's most critical issues. In 2014 water had ranked third among the most serious threats to business and society. The risks for water crises in the 2015 report were deemed both highly likely and highly devastating. The top ranking of water reflects the growing recognition among world leaders that diminishing supplies of reliable, clean water will be a real threat to health and wealth for the poor, for the richest economies and for the largest cities.

> One person out of seven has no electric power.

It is also notable that the WEF report reclassified water from an *environmental* risk to a *societal* risk. It has been recognised also by world leaders that nearly all human activity – food production, fishing, public health, industrial activities and power production – has water at its base.

Renewable energy technologies can make a major contribution to universal access to both energy and water in a sustainable way. In many regions with energy poverty there are abundant renewable energy sources. There is no lack of sunshine in sub-Saharan Africa or South Asia. In most regions of Africa there are more than 300 days of bright sunlight per year (Varadi *et al.*, 2018). Dry areas like the Sahara and the Sahel region can provide large areas with solar-powered electricity.

In rapidly growing peri-urban areas electric power grids may be available but need to be complemented with decentralised energy sources. Solar and wind can be part of new hybrid energy supplies. It is noted that there is a confluence of factors, such as greater urbanisation, population increase and economic development that will determine the energy mix. The United Nations (UN) Sustainable Development Goals of "clean water" and "energy for all" are strongly related and will depend to a large extent on solar PV and wind. This is further explained in Chapter 2.

> World Economic Forum: lack of water is both an environmental risk and a societal risk.

1.2 ACCESS TO CLEAN WATER

More than 650 million people lack clean water. Without safe water and sanitation people get caught in a vicious circle of poverty and sickness. In the poorest societies in the world, it is mostly women and children who lose precious time in their search for water and in the transportation thereof. Children die from diarrhoeal diseases that can be prevented. Open sewers running right through villages are far too common.

Water scarcity, poor water quality and inadequate sanitation have a significant impact on food security, educational prospects and other living conditions for poor families across the world. By 2050, the UN estimates that at least one in four people is likely to live in a country affected by chronic or recurring shortages of fresh water. The UN (www.un.org) summarises some of the huge challenges:

- 2.1 billion people lack access to safely managed drinking water services (WHO/UNICEF, 2017);
- 4.5 billion people lack safely managed sanitation services (WHO/UNICEF, 2017);
- 340,000 children under five die every year from diarrhoeal diseases (WHO/UNICEF, 2015);

- Water scarcity already affects four out of every ten people (WHO);
- 80% of wastewater flows back into the ecosystem without being treated or reused (UNESCO, 2017).

More than 650 million people lack clean water.

It is obvious that water supply and used water treatment need to be addressed simultaneously.

1.3 ACCESS TO ELECTRIC ENERGY

Today around 1,000 million people around the world lack electric power that could enable them to light their homes, cook their food or pump clean water (Jones & Olsson, 2017). Most of them live in rural areas of Africa and developing Asia. Another 1,000 million have unreliable electric power supplies (IEA, 2011). Furthermore, more than three billion rely on solid fuels and kerosene for access to cooking and heating (World Bank, 2017). The indoor and outdoor air pollution from burning wood and other biomass causes more than four million deaths each year.

Around 1,000 million people lack electric power. Another 1,000 million have unreliable electric power supply.

In sub-Saharan Africa alone, there are about 600 million people without access to electric power, around 57% of the population. Some 80% of these people live in rural areas. Less than 25% of the rural population have electric power. As a comparison 71% of urban residents in these countries have electricity (IEA, 2017b). Like lack of clean water, lack of electricity handcuffs poor families to poverty – especially women and girls, who must gather fuel and carry out the household chores. The good news, however, is that electrification efforts have accelerated so that electric power addition since 2014 is higher than the population growth.

The IEA (International Energy Agency) definition of access to electricity is at the household level and includes a minimum level of electricity consumption, ranging from 250 kWh per household per year in rural areas to 500 kWh in urban settings. The electricity supplied must be affordable and reliable. The initial level of electricity consumption should increase over time, in line with economic development and income levels, reflecting the use of additional energy services. (IEA/India, 2015).

It is well recognised that economic growth is closely related to access to energy. Electric energy consumption in Africa, particularly in sub-Saharan Africa, and in South Asia, is in disturbing contrast to the consumption in high-income countries. In sub-Saharan Africa the annual electric power production averaged 481 *kWh*/capita in 2012. This should be compared with the OECD average of 7,995 *kWh*/capita and the global average of 3,126 *kWh*/capita (Varadi *et al.*, 2018; World Bank, 2014). The contrast becomes even more upsetting when individual countries are compared, as in Figure 1.1. More details are found in Table 1.1 in Varadi *et al.* (2018), using data from the World Bank. sub-Saharan Africa (SSA) is the only region in the world where per capita access is falling (*ibid.*). Still it should be recognised that around 150 million sub-Saharan Africans have gained access to electricity since the year 2000 (IRENA (International Renewable Energy Agency), 2016d).

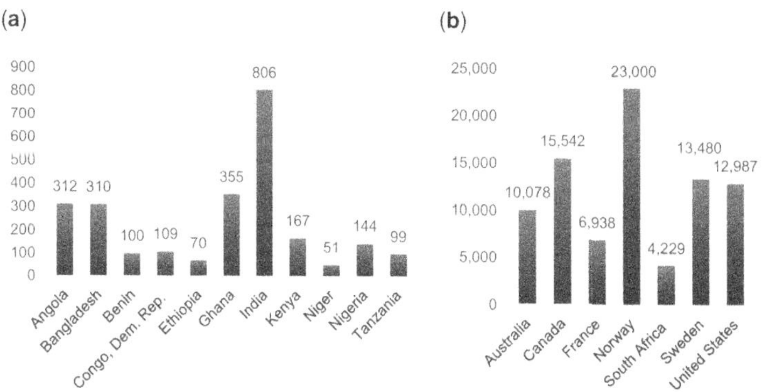

Figure 1.1 The annual electricity consumption (*kWh*/capita/year) in some countries. Low-income countries (a) are compared with high-income countries (b). Observe that the scaling to the left is more than an order of magnitude smaller than that on the right. *Note* the differences between Sweden, Norway, US and Canada. In the US and in Canada more natural gas and fossil fuels are used for energy production, while electric power is more widely used in Scandinavia. Data from World Bank (2014).

> The annual electric power production in sub-Saharan Africa is more than six times lower than the global average and almost 17 times lower than the OECD average.

The message from World Bank (2017) is clear: "In many countries with low levels of electrification access, both grid and off-grid solutions are vital for achieving universal electricity access – but they must be supported by an enabling environment with the right policies, institutions, strategic planning, regulations, and incentives."

Access to electric power is extremely important for access to clean water. For people living in remote and rural areas or in rapidly expanding peri-urban areas in poor regions of the world, power grids are out of reach. They cannot wait for conventional electric networks to be completed to solve their water supply or sanitary challenges.

1.4 DECOUPLING WATER FROM ENERGY WITH RENEWABLES

The close dependency between water, energy and food, the water-energy-food nexus, has been recognised for a long time. In my previous book (Olsson, 2015) the couplings and their consequences are described in detail. This book will show how renewable energy in combination with decentralised water operations can decouple many of these dependencies and meet challenges caused by climate change, population increase, water scarcity and poor water quality.

1.4.1 Renewable energy water footprint

Renewable energy can improve energy and water security. The energy sector relies heavily on water for energy extraction and production, accounting for 15% of water withdrawals globally. In a water-constrained world, conflicts with other end uses, such as agriculture, are intensifying and further impacted by climate change. With access to water increasingly recognised as a risk for energy security, it is becoming necessary to decouple energy sector expansion from water use.

Water is needed for fossil fuel extraction, transport and processing. Conventional thermal power plants, like nuclear, natural gas or coal, use huge amounts of water for cooling (Olsson, 2015, Chapter 13). Both water withdrawal and water consumption are significant.

The beauty of renewables is that they dramatically reduce not only the carbon footprint but also the water footprint. Solar PV and wind consume up to 200 times less water than conventional options (IRENA, 2015b). Substantial water savings are already being realised. Solar PV has a very low water footprint since water is not used for electricity generation. The water requirement, estimated at 118 litres/*MWh*, comes from the manufacturing of the PV cells and maintenance of modules (WEC, 2016, Chapter 8). Wind energy is certainly a low-carbon source, and the turbines have no water requirement during operation.

> Solar PV and wind energy have a very low water footprint.

1.4.2 Small-scale renewables

We will address the key issues – clean water and energy for all – and show the enormous potential of renewable energy, made possible by the technical developments of recent years. Around the world, low-carbon renewable energy is emerging as the go-to-green growth and poverty reduction strategy. The development of inexpensive solar and wind technology is considered a potential alternative, providing an electricity infrastructure consisting of a network of local-grid clusters with distributed electricity generation. This makes it possible to become independent of long-distance, centralised power-delivery systems. The emphasis in this book is to demonstrate how decentralised power from renewable sources and decentralised water supply and used water treatment offer new possibilities and hope for the un-privileged people left outside the advanced systems of today.

Renewables offer viable, affordable and scalable solutions. They are at the core of any strategy to meet climate goals while supporting economic growth, welfare, domestic value creation and employment generation. The potential of renewables is there for every country to harness. A major advantage of solar PV is that there isn't any minimal or maximal project size; it can be scaled to match the user load size and type. Solar PV can be used to power systems from the very small in size up to residential systems and utility-scale projects, ranging from a few *kW* to several hundred *MW*. Solar PV is at this moment probably the most attractive option for mini-grids (see also 3.2) for small villages (REN21, 2017a).

Renewable energy delivered by solar photovoltaic (PV) or wind will have a profound impact on water delivery and water treatment over the next decades. Any domestic user will require two kinds of energy: electric energy for illumination, machines, pumps, water treatment units and other equipment and thermal uses of solar energy for (1) process heat (including cooking), (2) ambient comfort, depending on the actual location.

We will describe how three kinds of renewables can satisfy these needs: solar photovoltaic (PV), wind power and solar heating. It should be noted that the direct use of solar energy for cooling via thermal processes has been tested in several situations, but so far it does not work satisfactorily.

Biogas is an important and environmentally friendly source of energy, and many rural areas depend on it. In this book we refer to biogas as a by-product of used water treatment but will not specifically consider the production of biogas as an energy source. We purposefully exclude some energy technologies, like geothermal energy, concentrating solar thermal energy and large-scale wind energy. In fact, the growth of bioenergy, concentrating solar energy and geothermal energy represented only 4% of renewable energy capacity growth in 2016 (IEA, 2017b).

We will show the potential of small-scale solutions that realistically can be used by individual households as well as small villages or a subdivisions of a city. Naturally, small-scale hydro can be considered an alternative energy source. This is already practised in regions with water resources but is not available in water-scarce areas. Regions with water scarcity or insufficient sanitation are certainly not areas where hydropower is an alternative. For the ongoing discussion we will exclude hydropower since our focus will be towards regions with water scarcity.

1.4.3 Providing water using renewables

Water is the critical element for a decent life and sustainable growth. Once the very basic needs connected to electric power are met, such as lighting and low-load production, new avenues open up. Renewable energy technology is already used to meet energy demand in many parts of the water cycle. Solar pumps can provide water for drinking, crop irrigation, increase access to piped water and reduce vulnerability to erratic rainfall patterns, thus increasing yields and incomes.

Renewable energy can also meet energy needs across the water supply chain, including various kinds of treatment such as desalination, water reuse and treatment, thus directly contributing towards access to both water and energy.

An important aspect is that the solar PV system has free "fuel" from the sun, while conventional fuels represent a major share of the operating cost. In many regions in rural Africa and developing Asia there are abundant solar resources. Even taking into account that the energy cost of desalination is relatively high, it is already acknowledged that solar-powered reverse osmosis desalination can produce water at a lower cost than fossil fuels. Likewise, wind power has free "fuel" from the wind. In each individual case it will be determined if wind is a viable complement or replacement for solar PV.

> With free "fuel" as much energy as possible should be extracted for good use.

When fuel is free the concept of energy-saving will get another meaning. Having free "fuel" means that as much energy as possible should be extracted for good use. The constraint is the power that will limit the number of appliances, water supply or water cleaning capacity.

1.4.4 Renewables versus nuclear and fossil energy

The interesting aspect of solar PV and wind power is that they are *technologies* and not *fuels*. They are unlimited, and the price will decrease as deployment increases. For fossil fuels it is the opposite: the more they are used, the more expensive they become (Wesoff & Lacey, 2017). Of course it should be remembered that fossil fuels have enabled our economy to develop. The message of today is that now there are realistic alternatives for producing energy.

The International Atomic Energy Agency (IAEA) has released the 2017 edition of its *International Status and Prospects for Nuclear Power* report series (IAEA, 2017). It states that the share of nuclear power in total global electricity generation has decreased for ten years in a row, to 10.5% in 2015, yet "this still corresponds to nearly a third of the world's low carbon electricity production." This means that renewables (including hydro, solar and wind) generate more than twice as much electricity (24.5%) as nuclear power and the gap is growing rapidly.

It is predicted that by 2022 renewables will be generating three times as much electricity as nuclear reactors. The International Energy Agency (IEA – not to be confused with the IAEA) in 2017 released a five-year global forecast for renewables, predicting capacity growth of 43% (920 *GW*) by 2022. The latest forecast is a "significant upwards revision" from last year's forecast, largely driven by expected solar power growth in China and India (IEA, 2017b). Non-hydro renewable electricity generation has grown eightfold over the past decade and will probably surpass nuclear by 2022, or shortly thereafter.

Globally, investments in fossil fuels will decrease to less than half of today's around $3.4 \cdot 10^{12}$ USD/year to $1.5 \cdot 10^{12}$ USD/year in 2050, while non-fossil energy expenditures show the reverse trend, increasing fivefold from around $0.5 \cdot 10^{12}$ USD/year today to $2.7 \cdot 10^{12}$ USD/year in 2050 (DNV GL, 2017).

Shifting investments to renewables, where the investment is up-front capital expenditures (capex), implies a shift from an energy system with a 60/40 split between operating expenditures (opex) and capex to one with the inverse split of 40% opex and 60% capex. In dollar terms, global opex will decline from about $2 \cdot 10^{12}$ USD/year in 2015 to $1.5 \cdot 10^{12}$ USD/year in 2050. Conversely, capital expenditure will increase almost 50% from $1.8 \cdot 10^{12}$ USD/year in 2015 to $2.6 \cdot 10^{12}$ USD/year in 2050. These figures do not include the cost of grids and energy efficiency investments (*ibid.*)

Eliminating the use of oil and gas will cut about 13% from the world's energy budget because mining, transporting and refining those fuels are all energy-intensive activities (Solutions Project, a US-based non-profit organisation). The greater efficiency of electric motors versus internal combustion engines could reduce global energy demand by another 23%.

1.4.5 Electric power cost development

Some recent reports emphasise that the energy field has undergone a massive change in less than a decade. Obviously we need tools to evaluate and compare costs for both conventional and renewable sources of energy, otherwise this could easily become an exercise in comparing apples and oranges.

An economic instrument is the levelised cost of energy (LCOE), which is defined as a way to express the lifetime costs divided by energy production and can be expressed in cost/*kWh*. The LCOE shows both capital costs in form of annual amortisations and variable costs. The LCOE is a step forward in the definition of a metric for the real costs of energy, though it also has its limits. The LCOE depends on the selected amortisation period and the reference interest rate. This is further examined in Chapter 12.1.

In addition, the LCOE doesn't say anything about the demand for power. A solar *kWh* in bright sun during summer has a different value from the same *kWh* in a cold region in winter, the same as a litre of fresh water has a different value in a hot desert from on the shore of a Nordic lake. It represents, however, a step forward in the comparison of quantities that by their nature are difficult to relate to each other.

A key observation from Lazard's latest levelised cost of energy (LCOE) analysis (Lazard, 2017) published in November 2017 is: "as LCOE values for alternative energy technologies continue to decline, in some scenarios the full-lifecycle costs of building and operating renewables-based projects have dropped below the operating costs alone of conventional generation technologies such as coal or nuclear. This is expected to lead to ongoing and significant deployment of alternative energy capacity." The report further notes that the global costs of renewable energy generation continue to decline. The LCOE for both large-scale solar PV and onshore wind technologies declined around 6% in 2017.

> The energy field has undergone a massive change in less than a decade. Solar PV and wind are now the electric energy sources with the lowest cost.

It is also observed that the gap between the costs of alternative energy technologies like large-scale solar PV and onshore wind energy compared to conventional generation technologies continues to widen. For example, the cost development for coal generation remains flat, while for nuclear power it is increasing. The LCOE for nuclear generation has climbed around 35% compared to previous estimates. The reason is the increased capital costs at various nuclear facilities in development. In 2017 new nuclear capacity of 3.3 *GW* was outweighed by lost capacity of 4.6 *GW* (Green, 2018).

It should be emphasised that the conclusions consider the global average. The actual cost will vary significantly from country to country depending on the cost of coal, gas, the cost of capital and the nature of the wind and solar resources.

Lazard (2017) emphasises the dramatic fall in the cost of large-scale solar PV from an average cost of 359 USD/*MWh* in 2009 to 50 USD/*MWh* in 2017, an 86% decline. This is half the cost of coal generation (102 USD/*MWh*). The cost of solar PV is far cheaper in some solar-rich countries. Contracts have been written at around 21 USD/*MWh* in Chile and 30 USD/*MWh* in Abu Dhabi (IEA, 2017a).

In the same period the global average LCOE for wind energy has fallen from 135 USD/*MWh* to 45 USD/*MWh*, a drop of 67%. On a global scale wind energy is today the cheapest electric energy. The global average LCOE (November 2017) for the most common energy sources is displayed in Figure 1.2.

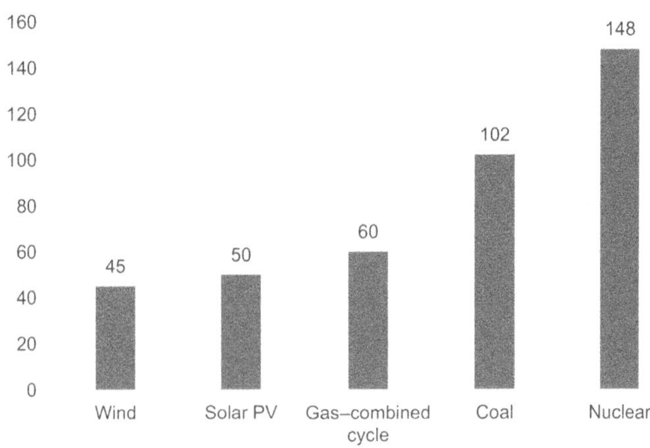

Figure 1.2 Global average of LCOE (USD/*MWh*) values in late 2017. Data source: Lazard (2017).

1.5 CLIMATE CHANGE CONSEQUENCES

A major and critical contribution of renewable energy is its impact on climate change, which is also one of the United Nations Sustainable Development Goals (see Chapter 2). The present global energy system contributes about 60% of all anthropogenic greenhouse gas emissions (IEA, 2017a). In particular, generation of electric power contributes over 40% of combustion-related CO_2 emissions. The carbon footprint varies vastly between different technologies and depends on power plant type, components, fuel types and waste intensity.

The combination of improved energy efficiency and renewable energy can give the world a realistic chance of limiting global warming to 2°C. At the same time, it will reduce air pollution, both locally and globally. This will have profound health effects from the public-health to the individual-household level.

Replacing fossil fuels with solar PV generation and wind turbines will reduce global CO_2 emissions.

Solar PV generation and wind turbines reduce global CO_2 emissions compared to fossil fuel energy production. The estimates of total global warming emissions depend on several factors. For solar PV it depends on the total solar irradiation and the number of sunshine hours. For wind turbines it depends on the wind speed and percentage of time the wind is blowing. The material composition of the solar cells or the wind turbines also contributes to carbon emissions.

According to NREL (National Renewable Energy Laboratory) (2012a), solar PV will generate around 40 g CO_{2eq}/kWh. Most estimates of wind turbine lifecycle global warming emissions are in the range 9–18 g CO_{2eq}/kWh. As a comparison, estimates of lifecycle global warming emissions for natural-gas-generated electricity are between 270 and 900 g CO_{2eq}/kWh and estimates for coal-generated electricity are 600–1,600 g CO_{2eq}/kWh (IPCC (Intergovernmental Panel on Climate Change), 2011).

1.6 THE NEED FOR COOPERATION

To make clean water accessible it is necessary but not sufficient to engage electrical engineering professionals to supply the electricity.

Water professionals, key customers, financial institutions, health service providers and educators need to be involved and engaged in a collaborative effort.

A system for clean water is built up of many components that must work together and operate reliably. The customer should not need to be an expert in operation or maintenance but should be able to make the basic actions and manoeuvres to run the system. Chapters 4–14 aim to explain processes and systems that can produce clean water and renewable energy so that water professionals and power engineers will understand each other better. Financing is a key condition for success, so we will illustrate some financing approaches in Chapter 12.

1.7 OVERVIEW OF THE BOOK

It is illustrated in Chapter 2 that the development of renewable energy is a key contribution to achieving climate goals. This development can also improve public health in poor regions. Clean water is a key factor, but also air pollution-related health hazards can be limited when solid fuels and kerosene are replaced by solar PV and wind. The UN Sustainable Development Goals (SDGs) all closely depend on access to energy and to water.

The development of renewable energy is nothing less than a revolution and gives for the first time a realistic opportunity to make electricity accessible to all. The main obstacle is not technology but political will, financing and education, as described in Chapter 3.

In Part II (Chapters 4–7) we describe various water treatment technologies for small-scale operations. Pumping, discussed in Chapter 4, is a key operation in almost all water supply and treatment processes. Desalination and membrane technologies are expanding exceptionally fast and present new and realistic possibilities to clean contaminated groundwater or surface water, as illustrated in Chapter 5. In coastal areas seawater can be desalinated at a realistic cost. Most of the developing rural areas lacking electricity today are warm and rich in sunshine. This makes it possible to produce clean water with simple thermal technologies, using no electricity. This option is discussed in Chapter 6 and should be remembered as a complement to the processes supplied by electric power. Treatment of used water can be accomplished with many technologies, and a brief overview is given in Chapter 7.

Part III (Chapters 8–11) describes renewable energy systems. Solar PV has the potential to be the dominating technology in warm and sunny countries, as described in Chapter 8. Wind energy can be an interesting complement to solar PV in many regions, discussed in Chapter 9. Solar cells of course only produce energy during the daytime, but wind and solar together can provide a more reliable source of energy. Still, both solar and wind are intermittent sources of energy; therefore, some storage of energy is needed in order to produce energy when there is no sunlight or wind. This storage challenge is the topic of Chapter 10. Finally, all components in a system producing clean water using renewable energy requires a system that can manage all the units and coordinate the energy flow in the system. This is described in Chapter 11.

Part IV of the book "Applying renewable energy to water operations" (Chapters 12–14) concentrates on "soft" issues. Financing renewables in low-income areas is a crucial issue and may be the biggest obstacle to making clean energy and clean water accessible to all. This is illustrated in Chapter 12. Any electric power system needs a land area and this aspect is described in Chapter 13. A number of cases of water supply and treatment powered by renewable energy are described in Chapter 14.

Part V looks at the future; in Chapter 15 we dare to predict the unpredictable, with a look forward to the year 2030. If you read it in a few years' time it may look amusing.

1.8 FURTHER READING

There is a lot of official material about the water and energy situation in the world. The UN reports UN Water (2014) and UN WWDR (2014) describe the water situation in the world and Olsson (2015) explains the many connections between water and energy.

Access to electric power is described in World Bank (2017) and in IEA (2017a, 2017c). The two reports REN21 (2017a, 2017b) reveal a lot of information about renewable energy. The situation in Africa is particularly emphasised in IRENA (2016b, 2016d).

The books Varadi (2017) and Varadi *et al.* (2018) present excellent information for the non-specialist. The former book tells the story of the remarkable development of solar power systems and the latter shows in detail the consequences of the solar power revolution for countries in Africa and the Middle East.

Chapter 2

Towards sustainability goals

"A little less conversation, a little more action"

Erna Solberg, Prime Minister of Norway.

"Can we afford Civilisation?"

Mark Twain 1835–1910.

The true impact of renewable energy is greater than the sum of the energy services it can provide. Renewable energy can solve many of the negative environmental, health, social and political impacts associated with conventional forms of energy. The United Nations has presented 17 Sustainable Development Goals (SDG) to transform our world. Two of these (SDG6 and SDG7) are directly related to our topic: clean water and sanitation for all, and affordable clean energy for all. The water and energy goals influence virtually all the Sustainable Development Goals that the global community has defined. Access to clean water and clean energy is closely coupled to the development of human health and well-being, environmental health and security.

© IWA Publishing 2018. Clean Water Using Solar and Wind: Outside the Power Grid
Author: Gustaf Olsson
doi: 10.2166/9781780409443_0019

2.1 THE UN SUSTAINABLE DEVELOPMENT GOALS

The 17 UN Sustainable Development Goals (SDG) were adopted by the international community in 2015 as part of the 2030 agenda for sustainable development (UN WWDR, 2014). It should be recognised that there are a lot of interlinkages between the various SDGs. Therefore, it is important to adopt an integrated approach towards their implementation.

Sufficient energy and water will be needed to meet nearly all the development goals. SDG6 – clean water and sanitation – depends a lot on the availability of renewable energy, recognising that much of conventional energy generation today depends on the availability of water. SDG7 – access to affordable and reliable, sustainable and modern energy for all – depends strongly on the development of modern renewables like solar PV and wind. The strong links between SGD6 and SDG7 are increasingly recognised (Olsson, 2015; IRENA, 2017a). The energy-water-food nexus is a growing concern for decision makers globally (WEC, 2016). Pumping and water treatment by biological processes or desalination will increase the supply of clean water. Conventional electric power technologies, such as thermal power plants, consume large volumes of water for cooling, while solar PV and wind generation have negligible water consumption.

As already noted, solar PV and wind power consume up to 200 times less water than thermal power plants. Furthermore, renewable energy can provide clean water using pumping energy and various treatments including desalination. It should be recognised that the role of renewable energy and water solutions will directly and indirectly contribute to all the other 15 SDGs. This is further elaborated in Table 2.1.

> Making renewable energy and water solutions available will directly and indirectly contribute to all the other 15 UN Sustainable Development Goals.

Water supply using renewable energy is a key technology to meet the UN SDGs. A workshop was conducted in 2016 at the Massachusetts

Table 2.1 Important links between the UN Sustainable Development Goals SDG6 (clean water and sanitation) and SDG7 (affordable and clean energy) and the other 15 SDGs (partly adopted from IRENA, 2017a).

SDG	Links
1 – No poverty	To eliminate poverty requires both basic energy and clean water. Both can stimulate economic activity. Decentralised energy can save expenditure on fossil fuels, which has a huge impact on the daily life of poor people in many parts of the world. Clean water will naturally influence health conditions, which in turn will affect general living conditions.
2 – Zero hunger	Renewable energy has been used for pumping. Irrigation that is not wasteful will improve food production and reduce vulnerability to droughts. Renewable energy can also provide the necessary energy for food preservation and refrigeration, thus reducing food waste.
3 – Good health and well-being	Clean energy for cooking can reduce risks for respiratory diseases caused by indoor air pollution. Decentralised electric energy can support the operation of health clinics and hospitals in remote areas. Renewables for power – replacing fossil fuels and the transportation of them – will reduce health problems related to outdoor pollution.
4 – Quality education	Today many children cannot study after school because their home lacks electricity. Renewable energy for lighting will allow studies and reading after dark, which is a significant change of life conditions in areas formerly without electric power. Renewable power will also make time available that was used previously for fuel collection.
5 – Gender equality	Usually women and children have the burden of fuel and water collection. The use of traditional biomass for cooking has significant health effects on primarily women and children. Electric lights can increase safety and allow girls and women to attend meetings after dark.
8 – Decent work and economic growth	An increasing share of renewables will give new job opportunities. Better water provision for agriculture will support jobs in food production. Electric energy supply can give several new job opportunities at the village level.

(Continued)

Table 2.1 Important links between the UN Sustainable Development Goals SDG6 (clean water and sanitation) and SDG7 (affordable and clean energy) and the other 15 SDGs (partly adopted from IRENA, 2017a) (*Continued*).

SDG	Links
9 – Industry, innovation and infra-structure	Available local energy from renewables can create new business opportunities, both directly and indirectly.
10 – Reduced inequalities	Available local energy can reduce the costs of buying external energy via fossil fuels. The installation and maintenance of renewable energy sources will create new jobs. Easier access to water will reduce inequalities between consumers previously connected to piped water and those consumers forced to buy water from vendors.
11 –Sustainable cities and communities	Cities and particularly peri-urban areas can reduce the carbon footprint of energy supply. A decentralised water supply and treatment can increase not only availability of affordable water but also resilience.
12 – Responsible consumption and production	Renewable energy has the potential to make global energy supply cleaner and safer. Decentralised water systems can increase availability of clean water in remote and rural areas. All these systems could be produced, installed and operated in an environmentally and socially sustainable way.
13 – Climate action	Increasing energy efficiency in combination with a massive upscaling of renewable energy can make it realistic to limit global warming to 2°C. Climate change will cause water scarcity in many regions. Renewable energy generation will require less water than conventional systems. Solar pumping and water treatment can provide water and help to adapt to changing conditions.
14 – Life below water	Production and transportation of fossil fuels has had a catastrophic impact on groundwater sources, rivers and oceans. The impact on marine life has been significant, changing the marine ecology and causing great suffering for all those who depend on seafood. Furthermore, CO_2 emissions from fossil fuel burning have caused both warming and acidification of the oceans. Renewable technologies that will replace or reduce the consumption of fossil fuels can reduce the spills of coal, oil or gas extraction, refining and transportation via pipelines or tanker traffic.

Table 2.1 Important links between the UN Sustainable Development Goals SDG6 (clean water and sanitation) and SDG7 (affordable and clean energy) and the other 15 SDGs (partly adopted from IRENA, 2017a) (*Continued*).

SDG	Links
15 – Life on land	Conventional energy systems have a significant and negative impact on the environment. This can be avoided using well-designed renewable energy. It can replace fuel wood and charcoal in off-grid locations, thus avoiding too much forest degradation. With water more readily available both agriculture and wildlife can be supported much more.
16 – Peace, justice and strong institutions	Renewable energy not only can provide access to clean energy to those lacking energy but also make clean water available. This will decrease social and economic inequalities both within societies and between regions and countries. Unlike hydrocarbon fuels, electricity is not easily tradable between different world regions. It is therefore very important to assess renewable energy resource availability on a regional basis. Furthermore, it gives the final user much more control of the power generation.
17 – Partnerships for the goals	Both renewable energy and clean water will contribute to the broader goals of sustainable development. To succeed it will require local, inter-regional and global cooperation and partnerships.

Institute of Technology (MIT) in association with the Global Clean Water Desalination Alliance (Lienhard *et al.*, 2016). On top of presenting the current status of large- and small-scale desalination the participants were asked to value the most interesting options for desalination. Two metrics were used: (1) technology readiness level (TRL) and (2) impact. This is a methodology introduced by NASA in the 1970s. The TRL levels define up to ten levels of technology maturity, ranging from idea through basic and applied research to pilot and full commercial-scale implementation. The impact is graded at five levels. The 17 participants viewed the highest TRL and impact on two technologies with almost the same evaluation: reverse osmosis (RO) powered by either wind (mostly for large-scale) and solar PV. The workshop also concluded that "integration of desalination and renewable energy at a small scale can provide clean water in areas of transient or sustained water scarcity with limited or non-existent grids". Desalination is reviewed in Chapter 5 and solar PV and wind are further described in Chapters 8–10.

A word of caution may be expressed here. There is no mechanism to force governments or organisations to implement the SDGs. A key condition is how financing is going to be realised. Here it is crucial how international organisations like the World Bank, the International Monetary Fund and various development funds act. This is well described by the term Washington Consensus, first used in 1989 by English economist John Williamson (https://en.wikipedia.org/wiki/Washington_Consensus).

2.2 PUBLIC HEALTH, GENDER ISSUES AND EDUCATION

Renewable energy can provide the basic energy needs in homes. The World Health Organization (WHO) estimates that more than four million people die prematurely each year because of indoor air pollution, caused by cooking with traditional biomass and inefficient stoves. In addition, the use of kerosene lamps for lighting further contributes to health hazards and accident risks, such as burns.

> Every year more than four million people die prematurely because of indoor air pollution.

The positive health effects of using renewable energy should be recognised. Satisfying energy needs has also a gender perspective as the health hazards in traditional households mostly affect women. Furthermore, if available, electric power can bring water close to the home and will also improve quality of life for women and girls, who are mostly responsible for bringing home water. Modern renewable energy can also reduce or eliminate the time required to gather firewood. This reduces unsustainable biomass harvesting but it also gives time for, primarily, women and girls to pursue education or generate some income for the family. With renewables available, electric lighting can be provided at home and in schools, so education is profoundly influenced.

About 1,000 million people in the world depend on health facilities without any reliable electric power and readily available clean water. Many more people rely on facilities with unreliable access to both energy and clean water. Off-grid renewable energy can deliver affordable and clean energy to remote healthcare centres.

Access to electricity and clean water has a profound influence on education. There are strong links between education, energy, clean water and economic development. Jones *et al.* (2018) point out:

- In sub-Saharan Africa 90% of children attending primary schools have no access to electricity;
- Half of schools in Peru and a quarter of village classrooms in India have no electric power;
- One child in three – an estimated 188 million children – attend schools that have no lights, no running water, no refrigerators, no fans and no computers or printers.

As the authors point out, the dominating question is whether these children will be equipped with sufficient skill to help their country to grow economically. Electricity provides improved access to water. Results from Kenya clearly demonstrate the consequences: access to water means better sanitation. This in turn dramatically decreased absenteeism of both pupils and teachers due to waterborne diseases such as skin infections, typhoid and cholera (Jones *et al.*, 2018).

Jones *et al.* (2018) also note that 40–60% of the world's unfarmed arable land is in sub-Saharan Africa. A key issue is to train farmers in the use of more effective irrigation and cultivation.

Oil exploration, refining and distribution have a significant water footprint under normal operating conditions, in terms of both quantity and quality (Olsson, 2015, Chapter 11). This is further exacerbated as a result of the unprecedented impact on water resources from leakages, dredging, refining or accidents in exploration or transport. Therefore, the consequences for water resources from oil operations or oil accidents will be increasingly grave (Zabbey & Olsson, 2017). Population growth, increasing water use for agriculture and industrial activities and climate change all increase water scarcity. This will have huge social and economic consequences for large populations. Researchers have examined data collected by NASA's Gravity Recovery and Climate Experiment (GRACE) satellite mission during the period 2002 to 2016 to track freshwater trends worldwide. In the report, published in May 2018 (https://climate.nasa.gov/climate_resources/167/), the researchers claim that data shows access to fresh water will be the biggest challenge to humanity in the twenty-first century. The GRACE data made it possible for the researchers to track changes in freshwater resources around the world even in areas where local data has been scarce or unavailable.

2.3 FURTHER READING

The UN Sustainable Development Goals, presented by UNDP, are published in detail on the web page http://www.undp.org/content/dam/undp/library/corporate/brochure/SDGs_Booklet_Web_En.pdf.

The World Health Organization (WHO) is an important source of information for topics related to water, health and sanitation. The web page http://www.who.int/topics/water/en/ is an entry point to fact sheets, statistics, guidelines and other information.

The gender issue is observed by the UN programme Gender and Water (http://www.un.org/waterforlifedecade/gender.shtml), where "gender" refers to the different roles, rights and responsibilities of men and women and the relations between them.

Chapter 3

The renewable energy revolution

"I'd put my money on the sun and solar energy.
What a source of power!
I hope that we don't have to wait until oil and coal run out
before we tackle that.
I wish I had more years left!"

Thomas A. Edison (1847–1931).

The technical and economic development of renewable energy will have an impact on many different aspects related to electricity availability and climate change, as well as links to water, water supply and treatment. It is no longer out of reach for poor people or those living in remote regions. The development will also have an impact on many social and economic aspects like gender issues, health hazards and consumer power. In 3.1 we illustrate the global picture of renewable energy. The off-grid development is of particular interest here, as emphasised in 3.2. Scalability of renewable energy is a key property that will give the end user control of the production, Section 3.3. The phenomenal cost development of solar PV and wind is illustrated in 3.4. This is linked with the expansion of solar PV all over the world, as we see in Section 3.5. The expansion of wind power is discussed in 3.6. The radical development of

© IWA Publishing 2018. Clean Water Using Solar and Wind: Outside the Power Grid
Author: Gustaf Olsson
doi: 10.2166/9781780409443_0027

solar PV and wind already has geopolitical consequences, as illustrated in 3.7. The renewables have already created massive job opportunities, both in manufacturing and in assembling and mounting the renewable energy systems, as illustrated in 3.8.

3.1 THE GLOBAL PICTURE

There is a dramatic change under way in the energy sector. Already 173 countries have established targets for renewable energy (IRENA, 2017a). In 2015 154 GW of new energy capacity was added globally: 61% came from renewables, and 90% of the investments in renewables came from wind and solar power (IRENA, 2017a). In 2016 almost two-thirds of the net new global power capacity of almost 165 GW coming online was renewable, a new record year. The solar PV development, driven by sharp cost reductions and policy support, was a key factor in this development. In 2016 new solar PV capacity around the world grew by 50%, reaching 74 GW. Solar PV additions were larger than any other electric energy source in 2016, even surpassing coal (IEA, 2017b).

Most of the global power capacity coming online today is renewable.

Renewables have increased almost exponentially over the last decade. Figure 3.1 shows how renewables (mostly solar, wind and hydropower) have developed. Notably, solar and wind have increased remarkably. In the period 2017–22 it is expected that solar PV will have the largest growth of all (www.iea.org/renewables). The IEA (International Energy Agency) forecasts that the share of renewables in global power generation will reach 30% in 2022, up from 24% in 2016.

The growth rate of solar PV is exceptional compared to other electric energy sources, with an average growth rate of 41% between 2010 and 2015, admittedly from low levels (39 to 219 GW globally). This corresponds to 20% of all newly installed electric power capacity. During the same period wind offshore (outside the coast) has increased 30% per year and wind onshore (on land) almost 18% per year. Total wind power grew from 180 to 405 GW. As a comparison, hydropower grew 3.3% per year (IRENA, 2017a).

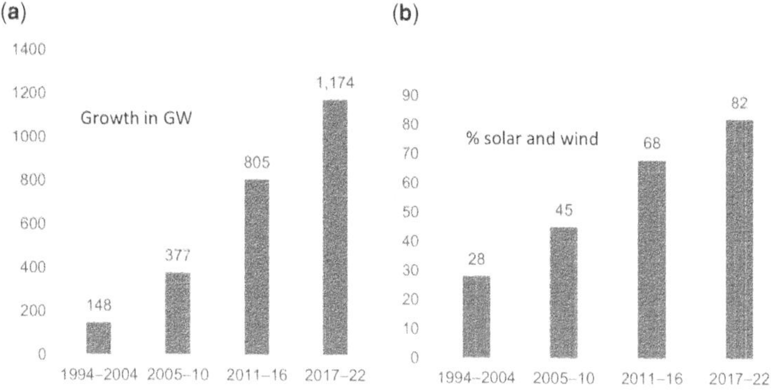

Figure 3.1 (a) Total renewable energy capacity growth (in *GW*) in four periods. The 2017–2022 growth is predicted. Most of the growth is in wind, solar and hydro. (b) The percentage of solar and wind in the total renewable power growth. Data source IEA (2017b).

China is the undisputed leader of renewable energy with more than 40% of the world's renewable capacity growth. One important driving force is the concern about air pollution. China surpassed its 2020 solar PV target in 2017, and it is expected that the wind target for 2020 will be reached in 2019 (IEA, 2017b). Chinese government agencies note that in the first half of 2017, renewables accounted for 70% of new capacity added (a sharp increase from 52% in 2016), thermal sources (mainly coal) 28% and nuclear just 2%. In late 2017 Beijing announced plans to stop or delay work on 95 *GW* of planned and under-construction coal-fired power plants, so the 70% renewables figure is set to see a healthy boost (Mathews & Huang, 2018).

3.2 OFF-GRID DEVELOPMENTS

Renewable electric power production in 2015 was dominated by large-scale generation (*MW*-scale and up). However, the market for small-scale generation outside existing power grids has been taking off. Bangladesh is the world leader in solar home systems. Small-scale renewables, mainly for lighting, are increasing rapidly in many developing countries, notably in Kenya, Uganda and Tanzania in Africa, China, India and Nepal in Asia, and Brazil and Guyana in Latin America (REN21, 2017a).

Renewable energy solutions, either off-grid or in mini-grid systems, are becoming economically feasible for many remote and rural areas (IRENA, 2015a; Varadi *et al.*, 2018, Chapter 3) where electric grid extension is not economically feasible. Considering the pace of grid-extension efforts globally it is likely that nearly 60% of the additional electricity needs to come from off-grid solutions, to attain the goal of electric power for all in 2030 (IRENA, 2017a).

Another reason for considering off-grid solutions is the relatively poor reliability of existing electric transmission lines. Around two-thirds of the sub-Saharan countries have transmission lines where at least half of the lines are more than 30 years old. Even in South Africa one-third of the transmission lines have been in operation more than 30 years. Aging infrastructure and lack of maintenance contribute to low reliability and common brown-outs or blackouts in sub-Saharan Africa (IRENA, 2016d). Furthermore, some systems are disrupted by wars and conflicts and it takes years to bring the systems back to full operation. Again, off-grid solutions can offer the end users opportunities to take control of their own energy supply.

As shown by Varadi *et al.* (2018), only a small fraction of people in Africa have access to the electrical grid.

Population growth in sub-Saharan countries is the biggest in the world. This, together with rising standards of living in Africa, means that a lot of new investments in electric energy are required just to keep a balance between demand and supply.

A mini-grid is defined as a structure of a size between an individual home system and a conventional power grid. Such a system may include a generation capacity in the range of around $1\,kW$ to the order of $10\,MW$ (sometimes power utilities call them "micro-grids" or "pico-grids"). They will supply electric power to several customers but operate in isolation from the national grid. In rural and remote areas, a mini-grid is often considered an attractive solution to provide lighting, water pumping and power for small production units.

Based on calculations made in 2010 by the IEA, UNDP (United Nations Development Programme) and UNIDO (United Nations Industrial Development Organization), the additional electric generation required to achieve electric power for all by 2030 is $468\,TWh$ for developing Asia, $463\,TWh$ in Africa and $10\,TWh$ in Latin America.

The electric power will be delivered via either stand-alone systems, mini-grids or national grids, as illustrated in Figure 3.2.

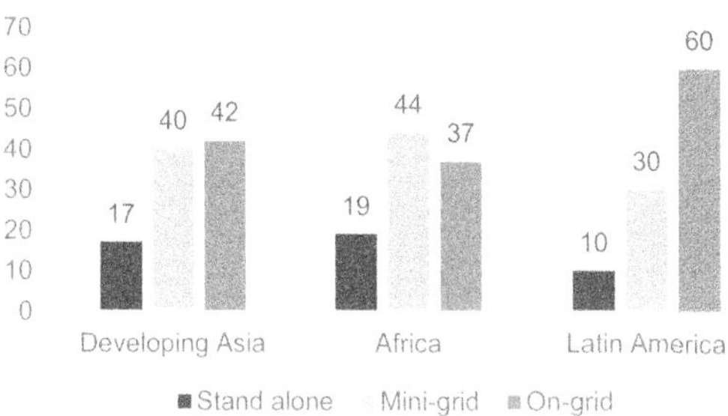

Figure 3.2 Structure of additional electric power supply (in %) to achieve electricity for all in 2030. It is apparent that most of the additional power will be added outside the traditional electric power grids (on-grid). Data from IEA, UNDP and UNIDO.

As noted in 3.1, solar PV and wind will dominate the development of renewable energy in the coming years. Solar is well above both wind and hydro. Already today almost 30 million people benefit from solar lighting products in Africa (Lighting Africa, 2013). The radical consequences for productivity, education, health services and quality of life are well documented. Almost 100 million people in developing countries have at least one solar PV lighting product in their home (BNEF and Lighting Global, 2016).

IEA made a forecast of renewables for developing Asia and sub-Saharan Africa in 2017 (IEA, 2017b). It is predicted that off-grid capacity will almost triple in the years 2017–2022, as shown in Figure 3.3. The off-grid growth is only a small share of the total PV capacity installed, but its socio-economic impact will be significant. Solar home systems (SHS) are expected to bring basic electricity to almost 70 million people in these regions over the five-year period. This will provide energy for lighting and small appliances, typically 20–100 W.

Low-income people can purchase this kind of service on pay-as-you-go schemes (see also Chapter 12.3). The energy cost is usually similar to or lower than the cost of traditional sources like kerosene lanterns. It should be noted that most SHS deliver DC power. Such a system needs no inverter (see glossary) to convert the DC power to AC but requires battery backup for use during the night. The DC system can supply power directly to DC-powered appliances like LED lights, radios, TV sets and mobile phone charging units. For larger power ratings, needed for water pumping and cleaning, AC power will be needed. Then DC/AC inverters will be needed; see Chapter 4.5.

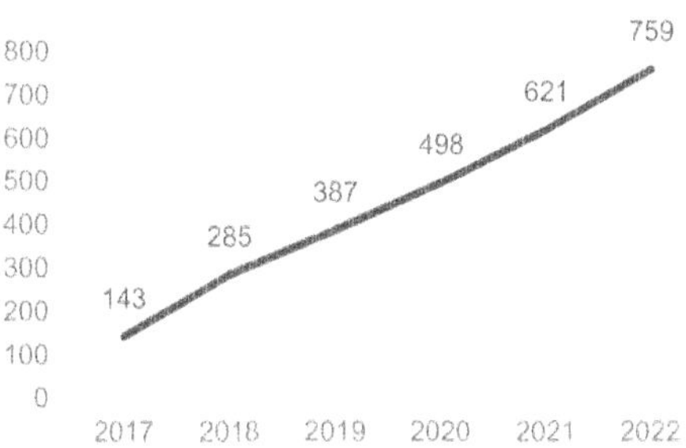

Figure 3.3 Predicted cumulative growth (in *MW*) of off-grid solar home systems in sub-Saharan Africa and developing Asia. The growth in mini-grids and industrial solar systems is of the same order of magnitude as the solar home systems. Data from IEA (2017b).

The progress of solar home systems is largely attributable to highly efficient LED lamps. Professors Isamu Akasaki, Hiroshi Amano and Shuji Nakamura made the first blue LEDs in the early 1990s. This enabled a new generation of bright, energy-efficient white lamps, as well as colour LED screens. As the Royal Swedish Academy of Sciences said when presenting the 2014 Nobel Prize in Physics to this trio of scientists: "The LED lamp holds great promise for increasing quality of life for over 1.5 billion people around the world who lack access to electricity grids. Due to low power requirements, it can be powered by cheap local solar power."

Another contribution to quality of life is the comparison of the light from solar-powered LED lights compared to the light quality from kerosene wick lamps. A LED light of 2 *W* will produce 380–400 lumens compared to 8–40 lumens for a kerosene wick lamp. The price per unit of useful light is anywhere between 3 and 100 times more expensive for the equivalent light from a kerosene lamp (IRENA, 2016d).

> Solar-powered LED lights provide much more useful light than kerosene lamps for the same cost.

The increased energy efficiency of lighting and home appliances can make larger off-grid solar home systems more affordable.

According to an interview with Adnan Amin, head of IRENA (Beckman, 2016):

I do believe that people are underestimating what is happening in off-grid. When we looked into this recently we discovered there is a lot of investment going into solar home systems, particularly in developing countries. You don't see this in the energy statistics; that's why we looked at the trade statistics. There are thousands of these home systems developed by entrepreneurs. They provide very low-cost basic power services for cell phone charging, refrigeration, and that sort of thing ... We need to improve the investment framework in developing countries. The aid model doesn't work. We need to incentivize entrepreneurs to start new businesses.

It is worth noting that electrification with renewables in remote areas does not necessarily depend on any national decision but on private initiatives. Of course, financing needs to be encouraged at all levels, including the national leadership.

> "People are underestimating what is happening in off-grid." Adnan Amin, head of IRENA.

The real driver of the renewable energy revolution is simply the reduction of costs for generating power. It is competitive with traditional fossil fuels and with a promise of even lower future costs. And even more important: the fact of diminishing carbon emissions gives hope for the future climate.

There is a huge potential for electrification of rural areas in Africa and developing Asia, as Figure 3.4 illustrates. Even in high-income countries off-grid solar is becoming more attractive. In countries and regions like Australia and California, and even parts of India, off-grid solar PV is becoming cheaper than purchasing the electric power from the power grid (WEC, 2016, Chapter 8). Naturally, being connected to the grid means a guarantee of electric power even if the off-grid systems break down.

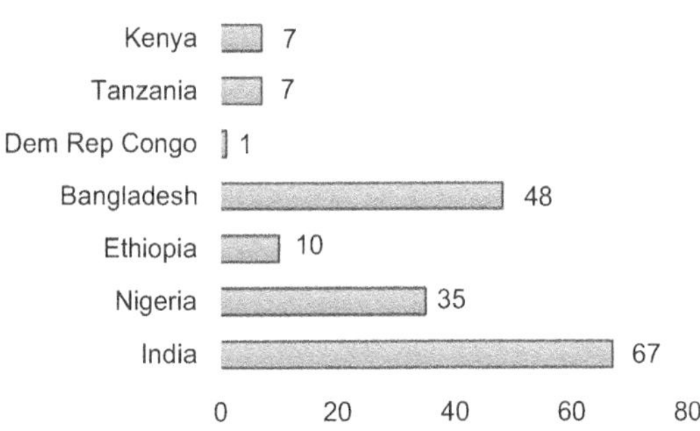

Figure 3.4 Rural electrification rates (%) in some developing countries. Data from IEA (2013).

It ought to be emphasised that increased electrification and rising electric energy use from renewable sources will not increase global warming. The production of CO_2 from fossil fuels must be decreased. So, any replacement of diesel generators and wood burning will make a positive contribution to the climate.

3.3 SCALABILITY OF RENEWABLE ENERGY

The scalability of renewables is remarkable, in particular for solar PV. The systems can be highly modular and can provide options for both off-grid and on-grid solutions. The scale can vary from very small lighting systems in remote areas to residential and commercial rooftop systems at utility scale. The size can vary from the *kW* range to hundreds of *MW*. In remote areas renewables can negate reliance on fossil fuels, like kerosene and diesel, which has not only an economic impact but also an effect on health.

A solar home system will start to fulfil basic needs such as lighting and low-load appliances. Adding more electric production capacity will allow power for water pumping, irrigation and treatment for water supply and water reuse.

Power delivery can be considered from three perspectives:

- Complexity,
- Hazardous and risky operation, and
- Customer power.

Solar PV systems can be from 1 to 100 to 1,000 solar panels. If the power capacity is too low, then new panels can be added; and conversely, if capacity is unnecessarily high, panels can be moved to another consumer.

An interesting feature of solar PV is that the panels are replicated, which means that the complexity of the system does not increase with the size, and its efficiency is not sacrificed with the power rating. This is not true for thermal power plants, which have an optimal operating size.

The scalability of renewable systems is remarkable, in particular that of solar PV. The systems can be highly modular.

In general, solar-powered systems do not present any hazards due to materials or operations. However, some chemicals are reason for concern, such as cadmium in Cd-Te solar cells and the components of most types of batteries.

The operational risk is yet another important factor. As renewables are scaled up they do not present increasingly greater hazards. Conventional methods of thermal power generation, nuclear or coal, have a typical optimum size for best efficiency. Usually this is quite a large power station, which also means greater system complexity. It is not profitable to build a 50 MW nuclear reactor. Renewables, on the other hand, are relatively benign technologies, without serious risks.

Scalability increases consumers' power. With local access to energy the individual citizen is empowered and is no longer dependent on a central authority. The promise of energy and water decentralisation is a democratic force that should not be underestimated. It is possible to start with a small investment in power generation. When there is demand for increased power generation, then more modules can be added. An important issue is project lead times; solar PV has the shortest of any power-generation technology. Furthermore, the mounting of solar PV panels does not require sophisticated equipment or highly trained personnel. Scaling up the power capacity in the future is also possible.

3.4 COST DEVELOPMENT OF SOLAR PV AND WIND

The falling cost of renewable energy technologies, notably solar and wind power, contributes considerably to the growing competitiveness of renewables versus conventional fossil fuels, as shown in Figure 1.2. Already today renewables like onshore wind, biomass, geothermal and hydropower energy are all competitive with or cheaper than any fossil fuel power stations, even without subsidies and despite relatively low oil prices (IRENA, 2017a). For obvious reasons, offshore wind and small-scale rooftop solar are about twice as expensive as onshore wind and utility-scale PV.

Costs for renewable energy technologies have fallen dramatically in recent years. The cost of solar PV modules decreased by 80% between 2009 and 2016. The prices for wind turbines declined by some 40% over the same period (IRENA, 2017a). This does not include the costs of balancing the grid system or compensating for the intermittent production. This will be the next economic challenge for solar and wind systems. As well as the solar panels or wind turbines the cost includes electrical system costs, costs for permits and installation. In an off-grid system the cost for energy storage should be taken into consideration (see also Chapter 10).

If the high social costs of pollution from extracting and burning natural gas or coal are taken into consideration, then renewables are even more competitive. This is demonstrated in China, where wind power already exceeded the contribution from nuclear power in terms of capacity in 2009 and in terms of electric energy generated in 2012 (Mathews, 2016).

> Traditional cost estimates for electric energy do not consider the high social costs of pollution from extracting and burning coal or natural gas.

There have been auctions to implement renewable energy in many countries. At the end of 2016, 67 countries had arranged such auctions and record low prices had been achieved for both solar PV and wind. For example, in Morocco a median price for solar PV of 30 USD per MW (or 0.03 USD/kW) was achieved, while a record low of 24.2 USD/MW (or 0.024 USD/kW) was offered for 300 MW in Abu Dhabi (IRENA, 2017a). This also means that there have been record low *energy* prices in auctions, as low as 30 USD/MWh or 0.03 USD/kWh (IEA, 2017a).

Solar PV is the most realistic and least expensive power source, and not only for remote areas outside the electric grids (compare Figure 1.2). Already today, in several countries electricity from small-scale distributed PV is cheaper than power from the grid. Solar PV has become cheaper than diesel-fuelled generation or lighting provided by kerosene. At the same time, it saves the user from the social and environmental problems associated with fossil fuels, including the transportation costs to provide them.

The cost of wind energy is also competitive with conventional power production (Figure 1.2). For example, in New Zealand in 2011 the long-term marginal cost for existing wind farms was in the range of 56–75 USD per MWh. The comparable electric energy price in the same period was 53–60 USD per MWh. Since then the cost of wind power has declined much more (WEC, 2016, Chapter 10).

Several experts were interviewed concerning the future of renewables compared to fossil fuels (REN21, 2017b). One question was whether or not the cost for renewables would continue to fall and will outpace all fossil fuels within the next ten years. Among the experts 67% agreed or strongly agreed with this statement and only 13% disagreed, while 20% neither agreed nor disagreed. The experts were quite uncertain about future oil prices. In contrast, future renewable energy costs were considered predictable and certain.

Even if solar PV and wind energy are already cheaper than diesel generator energy, it will be a challenging task to replace traditional generators in many places. The reason is that diesel generators have long been the technology of choice in off-grid areas. The diesel generator market and its supply chain are mature. In 2015, developing countries bought and installed some 600,000 units with a total capacity of 29 GW. About half of the power comes from units smaller than 0.3 MW (Climatescope, 2017). Furthermore, the carbon footprint from diesel-generated power is about twice as high as the average footprint from grid-connected power in Africa.

Despite the low cost of solar PV there is still an unfavourable competition with diesel generators. One important reason is lack of financing possibilities (see also Chapter 12). Diesel-generated power is two to three times more expensive than solar PV and the fuel cost is a significant contributor. For solar PV the up-front cost is higher while the fuel is free. So, even if diesel-generating power is more expensive to operate in the long run, diesel generators are cheap to buy.

3.5 SOLAR PV GLOBAL EXPANSION

The energy available from solar radiation is huge. The global average solar radiation received by the average m^2 can in one year produce the same amount of energy as a barrel (159 litres) of oil, 200 kg of coal or 140 m^3 of natural gas.

The solar energy obtained directly from the sun is termed solar radiation. There are two types of solar energy technologies: photovoltaic (PV) and thermal collectors. A PV cell will produce electricity directly from the solar radiation. Solar thermal collectors have been used for a long time for domestic heating and for providing hot water. The term solar *irradiance* is the measure of *power* from the sun over a certain area, typically measured in W/m^2. Solar insolation is a measure of the energy from the sun and is averaged over a long time, and expressed in Ws/m^2 per day or kWh/m^2 per year. The use of the two terms across literature is not fully consistent.

Global installed capacity for solar-powered electricity has seen an exponential growth, reaching around 227 GW_p (gigawatt peak electric) at the end of 2015 (WEC, 2016, Chapter 8). At the end of 2016 320 GW_p had been installed (Fraunhofer, 2016). It produced 1% of all electricity used globally. Solar PV installations are certainly not distributed equally over the continents; they are led by China, India, Germany and the United States (IEA, 2016a). Figure 3.5 shows that major solar installations have been made in regions with relatively less in terms of solar resource (Europe and China), while potential in high-resource regions (Africa and the Middle East) remains untapped (WEC, 2016, Chapter 8). There is a huge potential for solar PV in Africa, developing Asia and South America, where sunshine is abundant. This is further illustrated in Chapter 8.

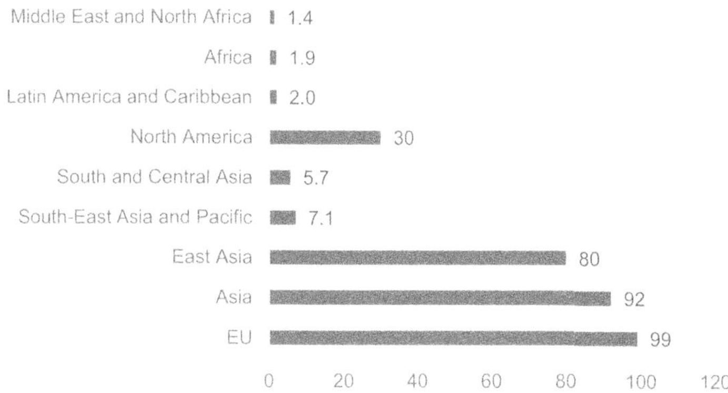

Figure 3.5 Solar data on installed capacity (*GW*) in different regions of the world at the end of 2015. Asia is shown both as a total and split into three major regions. Data source IEA (2016a).

3.6 WIND POWER GLOBAL EXPANSION

Global wind generation capacity at the end of 2015 reached 435 *GW*, which is around 7% of the world generation capacity (WEC 2016, Chapter 10). In just one year, 2015, the growth rate was more than 17%. Wind power capacity is estimated to increase to 977 *GW* by 2030. Most of the wind power (905 *GW*) will be onshore while 72 *GW* is expected to be offshore (*ibid.*).

Most of the wind power capacity in 2015, 422 *GW*, is from large onshore wind units, where the average machine size is 2 *MW*. Offshore wind, around 12 *GW*, is generated from 4,000 machines with an average size of 3 *MW*. Small onshore wind capacity makes a very minor contribution, less than 1 *GW*. The 800,000 small turbines produce an average of 1.25 *kW* (*ibid.*).

Naturally, wind resources vary between regions. According to IRENA (2016d) the northern, eastern and southern regions of Africa have excellent wind resources. This, however, is not reflected by the amount of installed wind turbine energy in these countries, as illustrated in Figure 3.6. The potential for harvesting wind energy ought to be enormous. The challenge is not the geographical conditions but the financial opportunities.

(a)   (b)

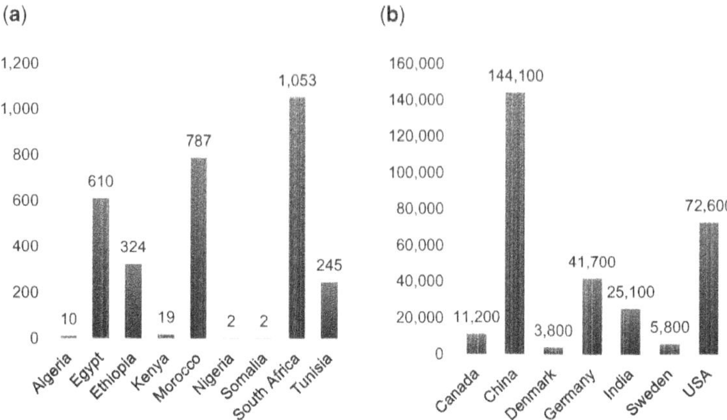

Figure 3.6 Installed onshore wind power (MW) in some African countries (a) compared to installed onshore wind power in some high-income countries (b). Note that the scales differ by two orders of magnitude. Data from WEC (2016), Table 11.

Wind capacity additions are led by China and the European Union. China's COP21 commitment indicated that wind capacity is to be expanded to $200\,GW$ by 2020 and solar power to $100\,GW$. There are already signs that China is pushing for higher targets for 2020: a possible 30–$50\,GW$ increase for both wind and solar PV (IEA, 2016a). This will almost surely have a global impact. Small-scale turbines are used for a variety of applications, including rural electrification and water pumping. They are installed increasingly to displace diesel generators in remote locations. In the five largest countries with small-scale turbines the upper capacity limit is between 15 and $100\,kW$ (WWEA, 2016).

The rapidly declining cost of competing technologies, such as solar, also poses challenges to small wind deployment. From these challenges, however, emerge innovation opportunities to increase the efficiency and reduce the costs of small wind technology.

Wind energy produced $226\,TWh$ in 2016, which avoided an estimated 160 million tons of CO_2 and could displace fossil fuel plant generation. This is equivalent to reducing the power sector's CO_2 emissions by 9% (AWEA, 2017). (The AWEA calculations are based on the EPA "AVoided Emissions and geneRation Tool", AVERT. AVERT calculates the pollution reductions provided by renewable energy and energy efficiency by statistically determining which fossil-fired power plants are most likely to have their output reduced due to the addition of renewable energy or energy efficiency). Also, land use is minimal since wind power generation can share the land with other activities (see also Chapter 13).

Due to the intermittent nature of both wind and solar PV (see Chapter 10), there are always issues of both lack of production and excess energy. DNV GL (2017) explains that there is too little economic consideration of what effect temporary surpluses of wind and solar power will have on the economics of renewables: *"we have not yet shown what the impact is of that."* However:

> *"we will also need to decarbonise heat, so it makes sense to store the surplus renewable energy, for example to heat water with it or to convert it into gas, to be used for heating. It is not yet clear what the most economic route will be."*

3.7 GEOPOLITICAL AND ECONOMIC IMPLICATIONS

Renewable energy has an impact on two important factors: economy and geopolitics. Renewable energy sources are products of manufacturing (Mathews & Tan, 2014). Using mass manufacturing means that the cost is related to economy of scale. Furthermore, as more experience is gained, manufacturing efficiency can be improved and can follow the scaling pattern characteristic of industrial learning curves. An increase of the production volume will almost surely imply a cost reduction. It seems to be a rule of thumb for a surprisingly big range of products that a doubling of the production volume leads to a 20% reduction in unit cost. There are surely other factors except production volume that will affect the price.

The fact that manufacturing can in principle be conducted anywhere means that renewables can offer genuine energy security. There is no geopolitical pressure where one country has deposits of a fossil fuel, but another does not. Renewables promise an end to energy security being closely related to geopolitics.

In Table 3.1 we list the top two countries in terms of total capacity or generation.

Table 3.1 The top two countries in terms of renewable energy total capacity or generation.

Type of Electric Power	1st Country	2nd Country
Renewable power excl. hydropower	China	United States
Renewable power capacity per capita (excl. hydropower)	Denmark	Germany
Solar PV capacity	China	Germany
Solar PV capacity per capita	Germany	Italy
Wind power capacity	China	United States
Wind power capacity per capita	Denmark	Sweden

Source: Data from REN21 (2017a).

<hr>

Renewables promise an end to energy security being closely related to geopolitics.

<hr>

Resources for energy are traded all over the world. This is dramatically illustrated on the website published by the British Royal Institute of International Affairs, known as Chatham House, https://resourcetrade. earth/. To generate traditional electricity requires international trade in coal, natural gas, petroleum and uranium. Now we see new competitors that cannot be traded internationally: wind and sunlight. Unlike commodity fuels, metals and other materials, which can be extracted in one place and transported for use somewhere else, including internationally, wind and sunlight are immaterial and are typically used where they are. Both wind power and solar power are nevertheless becoming steadily more significant contributors to electricity generation.

3.8 JOB SKILLS TO MOUNT AND OPERATE SMALL UNITS

The time needed to mount solar panels and additional equipment is very short compared to all other energy production types. While it takes several years to complete a large thermal power plant (nuclear- or coal-fired), a solar panel system for a small application can be finished in weeks, or even days. Most of the mounting work can be performed by non-specialist workers, who can mostly be recruited locally. This means that assembling renewable energy systems can offer both rural and urban employment.

To mount large wind turbines requires more special equipment and trained personnel. Naturally the need for advanced equipment is not as high for smaller wind turbines.

> Mounting of solar panels does not require highly skilled specialists or advanced equipment.

It is apparent that off-grid renewable energy deployment cannot be successful without technical assistance and human capacity-building. Education at different levels is a key ingredient. In a recent publication concerning electrification Dr Lawrence Jones, Vice President of International Programs at the Edison Electric Institute, suggested (Jones, 2018) that in order to achieve electric energy for all we need to think in terms of a metric expressed as "the number of engineers, technicians etc. per MW". Jones also points out that sub-Saharan Africa and south Asia need to substantially boost the number and capacity of engineers and technicians who can develop, install, operate and repair the components of both grid and off-grid equipment. An extensive analysis of job opportunities in the renewable energy industry in India is presented in CEEW (2017). Job opportunities in renewable energy are further examined in Chapter 12.2.

3.9 FURTHER READING

IEA and IRENA are key sources for a lot of information concerning renewable energy. IEA (2017b), IRENA (2016c, 2017a) and REN21 (2017a, 2017b) provide a lot of data. IRENA (2016a) contains statistical information on renewable energy for each country.

Solar home systems are described in detail in Lighting Africa (2013) and BNEF and Lighting Global (2016). The World Energy Council is a rich source of information on solar and wind development (WEC, 2016).

Varadi (2017) and Varadi *et al.* (2018) provide inside information and a broad overview of solar energy and its role in Africa and in the Middle East.

The newsletter *Energy Post* (http://energypost.eu/) is an excellent source of information about the rapidly changing arena of renewable energy.

There are several publications on mini-grids for rural areas. The National Rural Electric Cooperative Association (NRECA) has published a mini-grid design manual for rural areas (Inversin, 2000). A good guidebook for the connection of mini-grids (or pico-grids) smaller than 200 *kW* has been published by Greacen *et al.* (2013). A comprehensive report from USAID (2014) documents important lessons on rural mini-grids in developing countries. The challenges are a combination of technology, financing and organisation. A valuable record of seven case studies of micro-grids for rural electrification is documented in a UN report, prepared by Schnitzer *et al.* (2014). Five electrification cases from India and one each from Malaysia and Haiti are documented in detail. Experiences of mini-grids in rural electrification have been reported for Kenya (Kirubi *et al.*, 2009), Bangladesh (Yadoo & Cruickshank, 2010) and Nepal (Palit & Chaurey, 2011). It is essential to note that the technical issues are only part of the complete solution. Financing possibilities, high and unexpected expenses and lack of competence and of trust can cause systems to fail, even if the technical aspects are satisfactory (Cust *et al.*, 2007).

Part II

Water Technologies

Water operations need energy. The whole water cycle of water pumping and transport, treatment, consumption and collection and treatment of used water depends on energy. Therefore, the availability of local electrical energy is crucial in order to obtain clean water. Moreover, energy is essential to use water wisely and to reuse it whenever possible.

Pumping is part of many water operations and is the prerequisite for water transport. The energy aspects of pumping are discussed in Chapter 4. Different uses of surface water or groundwater require different water quality: water for irrigation can mostly be of lower quality than drinking water; grey water from washing can be reused for other purposes. Various water treatment methods are briefly described in Chapter 5. Electrical energy is not always required to obtain drinking water: solar still distillation is discussed in Chapter 6 and is a cheap and well-proven traditional method for obtaining fresh water. Used water must be treated and some traditional treatment technologies are described briefly in Chapter 7. The focus is on the energy requirement for the various technologies.

Chapter 4

Water supply

"Water is the driving force of all nature."

Leonardo da Vinci 1452–1519.

Too many people today still have to fetch water for their daily needs. Providing water via piping requires pumping capacity and electric power. Pumping is fundamental for all water supply and treatment as well as for irrigation. Some basic properties of pumping are examined in 4.1. It is important to consider many non-technical issues concerning pumping systems. In 4.2 a few of these are discussed in the context of developing regions. The parameters that characterise pumping are described in 4.3 and pump efficiency is defined in 4.4. Naturally there are several necessary components to a pumping system and some key parts are described in 4.5.

4.1 PUMPING

Pumping water – clean or contaminated – is a key operation in decentralised as well as centralised water operations. For the water supply the water must be moved from the source – a river or a lake – or drawn from underground to undergo treatment. In rural areas, water

© IWA Publishing 2018. Clean Water Using Solar and Wind: Outside the Power Grid
Author: Gustaf Olsson
doi: 10.2166/9781780409443_0047

pumping for irrigation is most essential. Pumping, for water supply, reuse and treatment of used water, is typically the major use of electric power in rural areas. Energy consumption is generally the largest cost in the lifecycle costs of a pump system, where pumps often run for more than 2,000 hours per year. To provide electric energy for pumping in rural areas of developing countries is not trivial. However, interesting products based on solar energy are now available.

For a small-scale water supply the water distribution pressure can be obtained either by a pump or by an elevated storage where the potential energy can provide the water supply pressure. If contaminated water is treated by desalination, there is a significant need for pumping energy; see 5.3 and 8.5. Any used water treatment or water reuse will use pumping energy (Chapter 7). It is apparent that pumping technology is an essential component of any water system, and having efficient pumping is crucial for any operation.

Among the advantages of solar PV pumping there are four often emphasised: unattended operation, low maintenance cost, easy installation and a long life. Both technology and economic viability have been considered in comprehensive literature reviews of solar pumping technology (Chandel *et al.*, 2017; Sontake & Kalamkar, 2016; Varadi *et al.*, 2018, Chapter 5.2). The authors have identified factors affecting performance of solar PV pumping systems and the degradation of PV modules as well as efficiency-improving techniques. It has been verified that solar pumping systems are more economically viable than diesel-based systems for irrigation and water supplies in rural, remote and urban regions (IRENA, 2016e). The investment payback time for some solar PV water-pumping systems has been found to be four to six years. This depends of course on local conditions, as shown below.

The costs of the systems are significantly different depending on their scale, purpose and configuration. It may be more meaningful to calculate the cost of the energy services provided and compare that to the existing costs that the user will pay for energy services off-grid.

4.2 PUMPING IN DEVELOPING REGIONS

The electric energy required for pumping is dramatically illustrated by the situation in India, where nearly 20% of electricity generation capacity is used for agricultural water pumping (CEA, 2016). India has around 26 million agriculture pumps, including at least 12 million grid-based electric pumps and ten million diesel-operated irrigation pump sets (IRENA, 2015b). Farmers pay only an estimated 13% of the true cost of electricity (Casey, 2013). The national burden of electric power subsidies is becoming too heavy. The subsidies encourage inefficient water use and contribute to depletion of groundwater. As water levels drop, more power is needed to pump the water, thus increasing the energy requirement of water extraction.

India has announced plans to replace many of its 26 million groundwater pumps for irrigation with solar pumps (Tweed, 2014). This will lead to large savings on installed electric power capacity and diesel and will hugely reduce CO_2 emissions. However, it is recognised that solar-based pumping poses a new risk for water resources: since the operational cost of solar PV pumps is negligible and the availability of energy is predictable, it could result in overdrawing of water. To combat that unintended consequence, the farmers who accept subsidies to purchase solar water pumps must switch to drip irrigation.

> When "fuel" is free it is tempting to overuse water for irrigation. Therefore, Indian farmers who accept subsidies to purchase solar water pumps must switch to drip irrigation.

In sub-Saharan Africa around 40% of the population, more than 300 million people, have no access to an improved source of drinking water from the region (UN Water, 2014). An analysis of data from 35 countries in sub-Saharan Africa (representing 84% of the region's population) shows significant differences between the poorest and richest fifths of the population in both rural and urban areas. More than 90% of the richest fifth in urban areas use improved water sources and over 60% have piped water on the premises. In rural areas, piped-in water is non-existent in the poorest 40% of households, and less than half of the population use any form of improved water source.

In the Sahel region solar-powered pumping stations have been in operation for almost two decades, providing better access to both electricity and water for two million people (IRENA, 2012). The region receives limited annual rainfall and the water table is at most 100 m

down. The energy to extract groundwater has helped the people to cope with the prolonged drought conditions. The population in the Sahel region of West Africa without access to safe drinking water dropped by 16% during a ten-year period leading up to 2009.

In Kenya only 6% of the agricultural land is irrigated and the main reason for this is lack of energy for pumping. There are some 2.9 million farmers in Kenya. An ongoing project, supported by the Renewable Energy and Energy Efficiency Partnership (REEEP), seeks to implement solar-powered irrigation (IRENA, 2015b). A typical system can pump up to 20 m^3 per day and operate at depths up to 15 m. The capital cost for the system is around 400 USD. Taking into account the savings on fossil fuels the payback time is estimated at two years. The programme aims for 30,000 pumping systems by 2018. There are positive social consequences: for example, women and children are relieved from manual pumping and carrying water. As in India, there is an apparent risk of groundwater overdrawing due to the negligible operational cost of PV pumps.

As noted by Varadi *et al.* (2018), Chapter 5.2, there are more than 30 solar water pump manufacturers in the world. The pumps can be purchased on the Internet, but systems are available locally in most countries in Africa and developing Asia.

4.3 PUMPING CHARACTERISTICS

The most common pump type is the *centrifugal pump*, in which the pump principle is to convert mechanical energy from the motor to kinetic energy (i.e., energy related to the fluid speed or flow rate) in the pumped medium, the water. This will create a pressure difference in the media between the pump inlet and outlet. Here we will skip many details of pump characteristics; more details can be found in Olsson (2015), Chapter 16.

The *system characteristics* or the *load characteristics* describe the pressure that the pump must produce to drive the flow. The pressure consists of both *static* and *dynamic* pressures. The static pressure (also called the static head) appears at zero flow rate and depends on how much the water must be lifted by the pump. In other words, a deep well requires a large static pressure to lift the water compared to a shallow well.

The dynamic pressure depends on the speed of the water in the pipe, in other words on the flow rate. The higher the speed the higher the dynamic pressure. If the pipe is wide, then the friction losses in the pipe are small and the dynamic pressure increases only slowly as the flow rate increases. Conversely, if the pipe is narrow then the water speed increases much faster when the flow rate increases. Consequently, the dynamic pressure will increase faster with the flow rate.

Actually, the dynamic pressure and the friction losses depend on the square of the water velocity v, in other words on the square of the flow rate Q. So, if the flow rate is doubled, then the power to provide the dynamic pressure must increase by a factor of four. The static and dynamic pressures are illustrated in Figure 4.1.

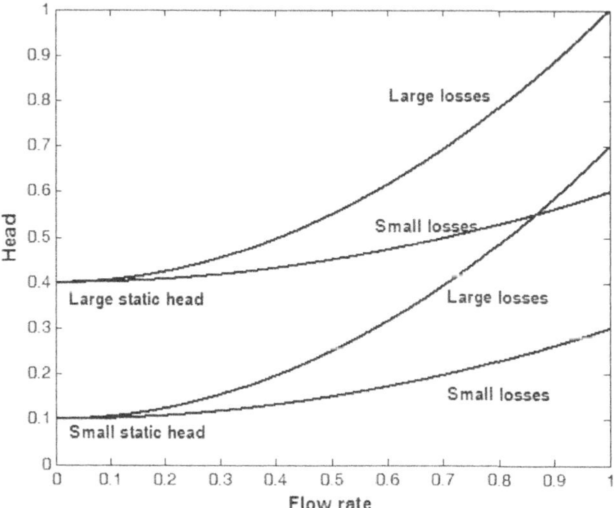

Figure 4.1 Different system curves that represent the load to the pump. If the water is lifted only a small height then the static head is small, and vice versa. If the pipe is narrow, then the losses are relatively high and the required dynamic pressure increases rapidly with the flow rate.

The pump curve, called the QH curve, describes the pressure that a pump can produce as a function of the flow rate, Figure 4.2. This is called the *pump characteristics*. The static and dynamic pressures are expressed as the *head*. Head is measured in metre liquid column and is proportional to the pressure that is created by the pump. Expressed differently: the head tells how high the pump can lift the water, given a certain pressure. The higher the flow rate the lower the head that can be produced by the pump.

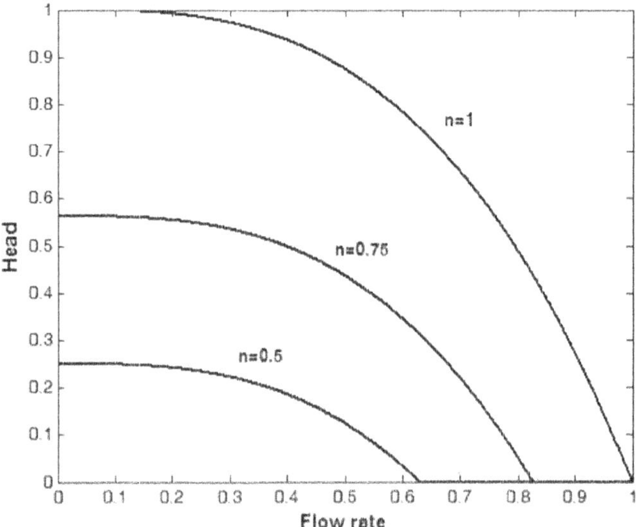

Figure 4.2 Typical pump characteristics or QH curves at different pump speeds (n). Q denotes the flow rate while *H* is the pressure, expressed in metres head.

The relation between the head (*H*, expressed in *m*) and the pressure (*p*, expressed in $Pa = N/m^2$) is

$$H = \frac{p}{\rho \cdot g} \tag{4.1}$$

Where ρ is the liquid density (kg/m^3), and g the acceleration of gravity (m/s^2). The formula indicates that pumping at a pressure of 1 *bar* (=10^5 *Pa*) corresponds to a water column (ρ = 1000 *kg/m³*) of 10.2 *m*.

If the rotational speed (*n*) of the pump is changed, then the QH curve is changed according to Figure 4.2. A lower speed means that the pump produces a lower head at a given flow rate, or produces a lower flow rate, given the head. If the pump is aimed to work at only one given head and flow rate, then the slope of the QH curve has no importance.

The operating point (or *duty point*) of a pump is determined by the intersection of the pump (QH curve) and system characteristics, as shown in Figure 4.3. The QH curve defines what the pump can produce, while the system curve (the load) defines which pressure is needed.

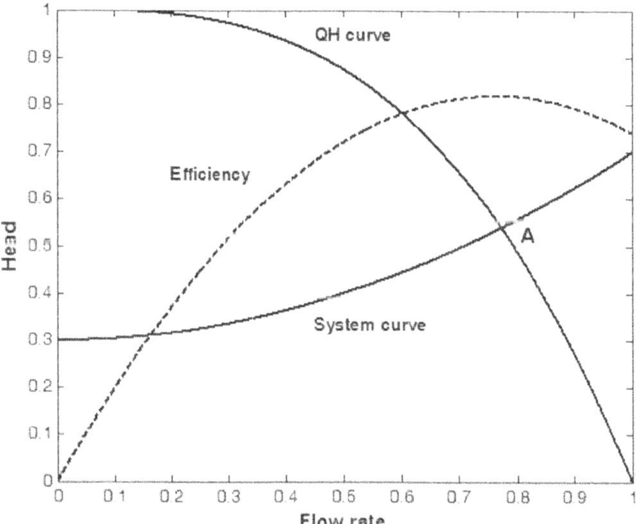

Figure 4.3 The duty point A of a pump is determined by the intersection of the QH curve and the system curve. The efficiency for a typical centrifugal pump is also shown. This indicates that the pump should be designed so that the maximum efficiency appears at the duty point. Expressed differently, the efficiency should be close to its maximum at the most common flow rates.

4.4 PUMP EFFICIENCY

The pumping efficiency depends on the flow rate as well as on the design of the pump.

- The electric power delivered to the electric motor connected to the pump is called the *incoming power.*
- The power transferred to the pump shaft is the *mechanical power* (the *shaft power*) and is slightly smaller than the incoming power and depends on power losses in the motor. The rated power of a motor is the mechanical power at the normal operating point and is slightly smaller than the consumed electric power. Typically, the motor efficiency is higher than 90%.
- The mechanical power is transferred to the *hydraulic power* P_{hydr} – the power that the pump transfers to the liquid in the shape of flow. This efficiency depends on the flow rate and is illustrated in Figure 4.3. The figure illustrates that the pump should be designed so that it has a maximum efficiency at the most common flow rates.

For the most common pump types, the term *power rating* normally refers to the *shaft* power and is measured in W or kW.

The efficiency of a large-scale pump may be around 89%. High-performance, small-scale pumps can reach an efficiency of 85%, but the performance of less expensive small pumps may be significantly lower.

The hydraulic *power* to lift water H metres is derived from:

$$P_{\text{hydr}} = Q \cdot H \cdot \rho \cdot g \tag{4.2}$$

where the hydraulic power is expressed in watts (W). Q = flow rate (m^3/s), H = head (m), ρ = liquid density (kg/m^3), and g = acceleration of gravity (m/s^2). Obviously, the required electric power must be higher than the hydraulic power. Naturally the total pump efficiencies depend on the actual equipment, and here we assume a typical value of 80%.

Example 4.1: *Power to Pump Groundwater*

Assume that the groundwater source is located 10 m below ground ($H = 10$). We need to supply 1 m^3 per hour ($Q = 1/3600\ m^3/s$). The water density is 1,000 kg/m^3 and $g = 9.81\ m/s^2$.

The required hydraulic power is:

$$P_{hydr} = \frac{1}{3600} \cdot 10 \cdot 1000 \cdot 9.81 = 27\ W$$

It is necessary to take some head loss in the piping system into consideration. We can assume this to be 10%. The required electric power, given an 80% motor/pump efficiency and a 10% head loss, then becomes

$$P_{el} = \frac{27}{0.8 \cdot 0.9} = 37.5 \approx 38\ W$$

The power is proportional to the head (depth) so a 20 m lift would need twice as much hydraulic power (54 W) and 76 W of electric power.

Doubling the flow rate also means doubling the required power. Thus, to lift 10 m^3 per hour from a depth of 100 m will require a hydraulic power of

$$P_{hydr} = \frac{10}{3600} \cdot 100 \cdot 1000 \cdot 9.81 = 2725\ W \approx 2.7\ kW$$

and around 2.7/(0.9 · 0.8) ≈ 3.8 kW of electric power.

To get the *energy* (in joules) the Q is replaced by the total volume V (m^3) in the calculations.

Example 4.2: Energy to Pump Groundwater

Let us calculate the energy to lift 1 m^3 of water from a depth of 10 m. This will require:

$$E_{hydr} = 1 \cdot 10 \cdot 1000 \cdot 9.81 = 9.8 \cdot 10^4\ J$$

where J is the energy measured in joules. With 10% pipe loss and 80% motor/pump efficiency taken into consideration the required electric energy input is around 13.6 · 10^4 J. Translating J (=Ws or watt seconds) to kWh as a more commonly used unit for electric energy:

$$1\ kWh = 1000\ (W) \cdot 3600\ (seconds) = 3.6 \cdot 10^6\ Ws$$
$$= 3.6 \cdot 10^6\ J = 3.6\ MJ$$

so, the electric energy of 13.6 · 10^4 J corresponds to 3.8 · 10^{-2} kWh. Another way to calculate the energy is to simply multiply the power (38 W) by time:

$$E_{el} = 38 \cdot 3600\ Ws = \frac{38}{1000}\ \frac{3600}{3600} = 0.038\ kWh$$

Again, the required energy is proportional to the head and to the volume of water. So, instead of lifting the water 10 m we now calculate for the head of 100 m and lift 10 m^3 instead of 1 m^3. This will require 100 times more (hydraulic) energy or $0.027 \cdot 100 = 2.7$ kWh; in other words, it demands the hydraulic power of 2.7 kW for one hour. The required electric energy for a pump system with 80% efficiency and pipe loss of 10% is 3.8 kWh.

If the pump is running during daylight solar hours (assuming eight hours) then the required energy is produced for eight hours (only 1/8 m^3 of water is pumped every hour), which will require the power $3.8/8 = 0.48$ kW from the solar PV system. This is 1/8 of the power calculated in Example 4.1.

In a solar PV pumping system, typically two types of system configurations are prevalent. In the first one a submersible pump lifts groundwater into an overhead tank that serves as an energy store and supplies the pressure needed for the pressurised irrigation system. In the second configuration there is no storage system and the water is pumped directly into the irrigation network.

Example 4.3: *Experiences from Solar-Powered Pumping*

Solar PV water-pumping systems are used for irrigation and drinking water in India (Roul, 2007). Most of the more than one million pumps in operation have the motor power 2.0–3.7 kW. Typically, a 1.8 kW_p (kW peak power, see also Chapter 8.2) solar PV array is used for irrigation purposes. Such a system can deliver around 140 m^3 of water per day from a total head of 10 m.

Let us compare this energy need with the theoretical hydraulic power (4.2). Assuming that the pump is working eight hours per day and only 60% of the peak solar power can be used:

$$P_{hydr} = \frac{140 \cdot 0.6}{8 \cdot 3600} \cdot 10 \cdot 1000 \cdot 9.81 W = 286\ W$$

Apparently, the PV/motor/pump systems are quite inefficient, or the solar array is designed with a large safety margin.

Example 4.4: *Rule of Thumb for Solar Water Pumping*

A common rule of thumb is that a 1000 W_p (1 kW_p) solar water pump can draw and pump around 40 m^3 of water per day from a source that is up to 10 m deep. Again using (4.2), we find that the hydraulic energy is:

$$Q_{hydr} = 40 \cdot 10 \cdot 1000 \cdot 9.81\,J \approx 4\,MJ \approx 1.1\,kWh$$

Assuming eight hours of sunshine, 80% motor-pump efficiency and 10% pipe loss the required electric power would be 0.19 kW. In other words, a large safety margin is assumed.

Typically, 40 m^3 of water per day is sufficient to irrigate up to one hectare of land planted with regular crops.

4.5 COMPONENTS IN A SOLAR PV PUMPING SYSTEM

The solar panels make up most of the cost in a solar PV pumping system, while the pump usually only represents a marginal cost. Of course, the size of the energy supply depends on the required flow rate (m^3/hour), and the solar irradiance.

4.5.1 Solar panels

One of the major advantages of solar flat panels is that they can produce economically interesting quantities of energy without the need to track the sun's position. The reduced demand for maintenance due to the lack of moving parts is reflected in a better overall economic result.

Typically, for most solar panels it is guaranteed that the module will produce 90% of its rated output for the first ten years and 80% of its rated output for up to 25 years. This is an outstanding operational life, longer than that for most other equipment. The limited complexity of mounting and operating a solar PV system (compare Chapter 8) and the safety of the solar cells in combination with the manufacturing costs are all factors in the success of the expansion of solar PV. Of course, if the equipment is in a remote area the commercial value of the guarantee may be quite limited. However, the fact that such a guarantee is still given is an indication of the reliability of the system.

4.5.2 Inverters and pump controllers

Solar panels deliver a direct current (DC) and the pump can be either a DC or an AC type. For an AC pump the DC from the solar panels must be converted from DC to AC. This is performed in an inverter, a power electronics device. A DC pump, however, does not need an inverter. DC pumps are more common for small-scale applications, typically less than 3 *kW*. This is adequate for a single household or for a small irrigation systems.

DC motors tend to have higher efficiency than corresponding AC motors. Traditionally it has been simpler to control the speed of a DC motor. Today, however, it is more economical to install an AC motor with an electronic variable speed controller. Furthermore, DC motors have many moving parts that are expensive to replace. They are supplied with brushes and commutators that have a limited operational life and must be replaced regularly. An AC motor with power electronics controller is both more reliable and more economical. It has no brushes and has a rugged design.

Inverters represent a proven technology over a very wide range of power levels and usually have very high efficiency: 98% or higher. Inverters for solar PV systems are often subject to harsh conditions, exemplified by operation during many sun hours combined with temperatures higher than 40°C. Dust is another challenge. When a small inverter is coupled directly to a PV module and mounted on the rear side the conditions may be severe. The temperature may reach up to 80°C. Under these conditions the electronics must be ready to operate for a period comparable with the lifetime of the PV modules, at least 20 years. Therefore, it is important to have a design margin that will ensure long-term reliability. The individual components must meet the harsh conditions. Given an adequate design the useful life of an inverter will exceed 20 years.

There is an internationally accepted standardisation of test procedures for PV system components. The standard IEC1215 defines the electrical and thermal properties of the components and can give an indication of the expected lifetime.

The pump also needs a controller that can adapt the power directed to the pump with the power delivered from the energy source. Generally, a controller should be provided with a voltage protection capability that will shut down the pump if the supplied voltage is too low or too high. In Chapter 11 we analyse more details concerning the control equipment.

4.6 FURTHER READING

Many pump manufacturers provide information about pumping principles and equipment. Grundfos Pump Handbook presents an excellent description of pumping principles (www.grundfos.com). The full link to their – now classic – work about pumps is http://machining. grundfos.com/media/16620/the_centrifugal_pump.pdf.

Solar array-based pumping technology is now commercially available. One example is the Lifelink concept from Grundfos (www. grundfos.com/market-areas/water/lifelink.html).

Another large pump manufacturer Xylem (https://www.xylem.com) has a web-based pump selection tool, Xylect, to guide the customer to find a suitable pump.

There is a lot of practical information on www.youtube.com. Search for "solar pumping systems".

Chapter 5

Water treatment

"We never know the worth of water until the well is dry."

Thomas Fuller, English churchman and historian (1608–1661).

The gap between available water resources and our domestic, municipal and industrial needs is becoming wider. Key causes are climate change and population growth. The number of people living in regions affected by severe water stress is expected to increase by a billion to almost four billion in the next two decades. The most severely affected regions will be North Africa, the Middle East, northern China, southern India, Pakistan and certain parts of the United States and Mexico. At the same time almost 40% of the global population live less than 100 *km* from the sea and 60% of the world's biggest cities in coastal areas do not have access to fresh water.

The available groundwater or surface water in remote regions may have a low quality and the contaminants need to be eliminated or reduced. The water must be cleaned to an acceptable quality, which depends on the final use. Naturally, drinking water needs to satisfy much higher quality standards than water for cleaning or for irrigation. In 5.1 an overview of water treatment technologies is presented. Filtering is

© IWA Publishing 2018. Clean Water Using Solar and Wind: Outside the Power Grid
Author: Gustaf Olsson
doi: 10.2166/9781780409443_0061

often the first step in water purification, Section 5.2. Desalination is a key technology to produce clean fresh water, Section 5.3. Disinfection, 5.4, is another important means of removing pathogens and protecting human health.

5.1 PRODUCING CLEAN WATER

The required energy to clean water depends on the quality of the raw water supply. Often, available water sources in developing countries are contaminated and consequently there is a high demand on energy use by volume of water (such as kWh/m^3) to treat the water to an acceptable quality.

5.1.1 Underground water resources

Groundwater is the major source of water in many parts of the developing world and there is a serious problem in that in many places the water table is declining. As mentioned in Section 4.2, underground water has been pumped in an unsustainable way in many regions. There are two consequences: water sources are dwindling and the pumping energy to reach them is increasing. In many areas water that has been pumped from aquifers has been underground for thousands of years and is now being used far in excess of the rate at which the water is replenished. It may take generations or centuries to refill these aquifers via rainwater. Such a resource is non-renewable and this kind of practice can be called mining of groundwater. Irrigation will of course influence groundwater resources. Since the irrigation is mostly done in dry or arid areas there is a large risk that the groundwater levels will decrease, since the groundwater will not recover as quickly as it is consumed.

5.1.2 Saline water

The need to use seawater as a source for fresh water has increased dramatically because of both population growth and climate change. Many dry regions have become drier. Using seawater as a source represents an important possibility in coastal areas. The salt concentration in seawater is typically 3.5% or 35,000 *ppm* (parts per million or *mg/l*). Many underground aquifers or other groundwater sources have brackish water that needs to be made drinkable by eliminating the salt. Another increasing water supply problem is the invasion of seawater into groundwater. Sometimes the groundwater resources have been exhausted. In other cases, the rising sea level caused by climate change will intensify the problem. Table 5.1 indicates typical salinity for various water qualities.

Table 5.1 Salinity for different kinds of waters.

	Salinity (ppm = mg/l)
Seawater	35,000
Highly saline water	10,000–35,000
Brackish water	1,000–10,000
Fresh water	<1,000

5.1.3 Contaminated water

It is crucial to make drinking water safe to drink. Many groundwater or surface water sources in remote areas have highly contaminated water, not only pollution from organic carbon (COD) or nutrients (ammonia and phosphorus) but also pathogenic bacteria, viruses, protozoa and worms. Many sources are highly contaminated due to the presence of harmful metals like fluoride and arsenic, which cause serious health hazards. Obviously, unsafe contaminants must be eliminated. The surface water sources are frequently used as dumping grounds and at the edges of the water bodies there is often defecation, which results in bacteriological contamination. This is a challenge not only to agricultural production and rural livelihoods, but also to food safety.

> Treating contaminated water requires more energy.

To remove contaminants requires energy. Biological treatment to remove COD typically requires a certain energy per *kg* of COD. Disinfection of water can be done in several ways. One principal way is disinfection via UV light. Another is via filters of various sizes. Both of them require electric energy to provide the UV light or to produce a pressure to operate membrane filters.

5.1.4 Water treatment technologies

The protection of human health is among the most important goals of water treatment systems. Any type of treatment must aim to reduce or inactivate potentially pathogenic organisms. Membrane separation is one methodology to remove harmful substances from the water. This aspect is emphasised in 5.2. Another technology to remove pathogens is by disinfection, as described in 5.4.

Having saline or brackish water as the water source means that small molecules have to be removed in order to make the water drinkable or useable for irrigation. This is called desalination, and is discussed in 5.3. There are two main technologies to produce fresh water from seawater or brackish water:

- Distillation, or thermal methods (heat treatment), and
- Reverse osmosis (membrane process).

Distillation is the oldest technology to obtain salt-free water. It uses a heat source to evaporate the water into steam and a cooling source to condense the steam into desalinated water. Regardless of the salt levels in the incoming water, the produced water will generally have a final salinity of less than 10 mg/l.

The phenomena of osmosis and reverse osmosis (RO) have been known for about 100 years but only in the 1960s, with the development of synthetic membranes, did these principles become an industrial reality. The first membranes were made from cellulose acetate. Since then a large number of organic membranes, made of polymers, and even mineral membranes have been added to the list. An RO membrane only allows water to pass through and retains the solutes. Membrane desalination uses high pressure from electrically powered pumps to separate fresh water from seawater or brackish water using a membrane. In other words: the RO process is electric power driven. Membrane technology, mainly reverse osmosis (RO), is used for almost 60% of installed capacity. Membrane technologies continue to dominate the desalination market. For example, according to the International Desalination Association (www.idadesal.org), in 2017 membrane technology accounted for 2.2 million m^3/day of annual contracted capacity while thermal processes accounted for just 0.1 million m^3/day during the same period.

5.2 MEMBRANE SEPARATION

Membrane technology has a huge impact on water purification. Semi-permeable membranes are used to physically separate substances. Using pressure across the membrane can drive the process. Then the smallest molecules or particles in a given solution are pushed through the membrane while larger molecules or particles are kept back. Pressure-driven membrane separation can be divided into four different types:

- *Microfiltration* (MF) screens particles from 0.1 to 0.5 microns (10^{-6} m);
- *Ultra-filtration* (UF) screens particles from 0.005 to 0.05 microns;
- *Nanofiltration* (NF) screens particles from $0.5 \cdot 10^{-3}$ to $1 \cdot 10^{-3}$ microns;
- *Reverse osmosis* (RO) ranging molecular size down to about 1 angstrom (10^{-4} microns). At this size the "particles" are individual molecules.

MF can remove suspended solids, high molecular weight species, bacteria, pathogens such as cryptosporidium and giardia in drinking water. Cryptosporidium is a parasite that commonly occurs in lakes and rivers, particularly when these water systems are contaminated with sewage or animal waste. The MF and UF techniques do not require any chemicals to inactivate the microbes.

Water purification by UF can remove macromolecules, colloids, viruses, proteins and pectins. The UF does not remove all the natural minerals, such as calcium (Ca^{2+}) or – more important – the salinity of seawater.

NF can remove small molecules and polyvalent ions such as calcium (Ca^{2+}) and magnesium (Mg^{2+}), while RO is needed to remove soluble salts, smaller ions, colour and low molecular weight species.

Another parameter that distinguishes the four types of membrane filtration from one another is the pressure under which they normally operate. The flux (the capacity of purified water, permeate, measured in litres per m^2 of membrane per hour) depends on the feed pressure. MF and UF need relatively low pressures, while NF and RO require much more. Typically, NF would need 1–4 *MPa* (10–40 bar), while RO would require 1.5–8 *MPa* (15–80 bar). Above the optimum pressure clogging of "pores" occurs and the membrane is compacted.

Membranes can be used to treat various kinds of contaminated water, like greywater, blackwater and urine. This is further examined in 7.2.

5.3 DESALINATION

Worldwide about 300 million people get some fresh water from more than 19,300 desalination plants in 150 countries. Middle Eastern countries have the highest investment in desalination, but the technology is increasingly used around the world in water-scarce regions. According to the International Desalination Association, around 87 million m^3 of fresh water is produced every day via desalination. If that amount of water could be equally shared between the world's 7.6 billion people, then everybody would have more than 11 litres every day. The annual growth of desalination is projected to be 12% over the five years from 2018 to 2022.

> Global desalination capacity today is sufficient to supply each person with 11 litres/day.

5.3.1 Energy supply for desalination

Today most of the energy supply for desalination is produced from fossil fuels and less than 1% comes from renewables (IEA-ETSAP & IRENA, 2013). Fossil-fuelled desalination has its problems, including the fact that electric power generated from coal and gas plants consumes water. Using a water-intensive resource to produce water is not sustainable. Therefore, we need to think about water as an energy resource and energy as a water resource.

The fact that fossil fuel for water production is not sustainable from an economic and environmental point of view has also been recognised in oil-rich Saudi Arabia, where King Abdullah's Initiative for Solar Water Desalination was announced in 2010 (IRENA, 2015b). The project has the goal of increasing water security for the country but will also contribute to the development of low-cost solar-based desalination technology. Increasing scale of deployment will make the solar desalination affordable in the long term. The cost of input energy is the dominating cost of desalination and is more than 50%. Considering continuously rising energy costs and with the impending exhaustion of conventional energy resources, it is a very attractive and promising prospect to use renewable energy to produce clean water. This is particularly true in the case of solar energy since regions with great water shortages tend to be those with higher solar radiation.

The water supply source for desalination is not always seawater, as shown in Table 5.2. Brackish groundwater is common in many water-scarce regions in the world.

Table 5.2 Potable water sources (globally) for desalination.

Desalination From	Percentage
Seawater	59
Brackish water	21
River water	9
Pure water for industrial applications	5
Used water for reuse	<5

Source: Wikipedia.

In many parts of the developing world there is plenty of both brackish water and solar energy. The use of solar energy to supply power for local desalination of brackish water is an interesting option where potable water is scarce.

5.3.2 Distillation – thermal methods

By heating up the water it is possible to separate the salt and dissolved minerals from the water in seawater or brackish water. This is called distillation. The basic principle of distillation is simply to heat up the water, let it evaporate and then condense it, producing fresh water. This principle has been used in warm and dry countries over the centuries by using sunlight. In ancient times sailors at sea obtained drinking water from primitive distillation systems. Inventors like Leonardo da Vinci and Thomas Jefferson experimented with seawater desalination. Aristotle, as early as the fourth century BC, described a method to evaporate impure water and then condense it to obtain potable water (Kalogirou, 2005).

Distillation requires a substantial amount of heat energy and must be coupled with other heat-producing applications. Distillation has mainly been developed in oil- and gas-producing countries, where it can be linked to thermal power facilities and thus use the heat they emit to produce evaporation. The advantage of this technique is that it does not require special pre-treatment of the water before its evaporation.

Today distillation is done more elaborately, and the most common method is *multistage flash distillation* (MSF) where the water is heated and the pressure decreased so that the water "flashes" into steam. MSF requires large amounts of energy to produce fresh water (typically 12–15 *kWh* and sometimes as much as 25 *kWh* per m^3). Two types of energy are required for the operation of a thermal desalination plant:

- Low-temperature heat, which is the main portion of energy input,
- Electricity, which is used to drive the system's pumps. Solar PV or wind power can be used to power the pumps. This may require 3–5 *kWh* per m^3.

The other thermal desalination process is *multiple effect distillation* (MED). Here the water passes through several evaporators in series. Vapour from one series is subsequently used to evaporate water in the next.

The MSF and MED technologies are industrial processes suitable for large-scale operations. They are very expensive for small-scale operations. MED is more efficient than MSF.

Thermal desalination is still the dominating technology in the Gulf countries and North Africa, but globally membrane-based methods are now the most common ways to desalinate seawater.

Distillation plants generate less waste (called brine) than membrane-based methods like reverse osmosis (RO), and there are no filters or membranes to get clogged. The brine issue is further examined below.

The energy used by desalination systems may be in the form of either work (such as electricity) or heat (normally as low-temperature steam). These forms of energy are distinct and cannot simply be added to find a "total" energy requirement. Because of the second law of thermodynamics electric energy can be converted to heat, but low-temperature heat cannot be converted to electric power.

5.3.3 Reverse osmosis

RO units are available in a wide range of capacities due to their modular design. The RO technology is used in applications from family size to as large as 600,000 m^3/day. This scalability together with similar scalability properties of solar PV and wind energy makes this combination extremely interesting.

Large plants are built up with hundreds of units that are accommodated in racks. A large plant's capacity far exceeds 100,000 m^3/day or 1 m^3/second, while very small units are made for flow rates as small as 0.1 m^3/day or about 4 litres/hour. The small units are used for single households, marine purposes or hotels and are suited to and used in rural areas or islands where there is no other water supply available.

Natural osmosis (or direct osmosis) governs how water transfers between solutions with different concentrations. It is the basis for the way in which human skin and organs function, and how flora and fauna maintain water balance. The osmosis process can be explained when there is a semi-permeable barrier such as a membrane located between two solutes with different salt content, as in Figure 5.1. In natural osmosis the water tends to flow from a solution with a lower concentration to a solution with a higher concentration if no external pressure is applied.

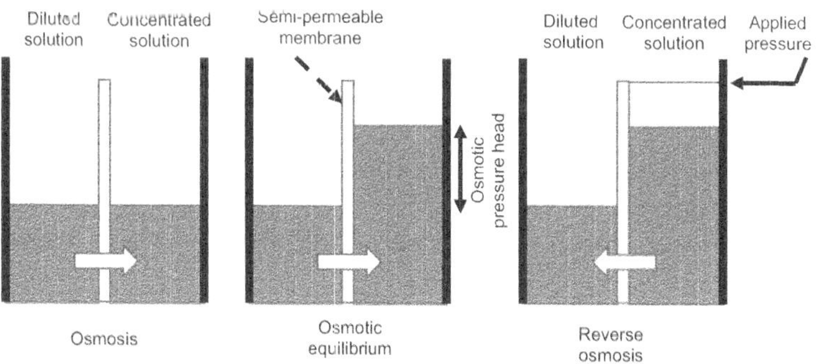

Figure 5.1 Illustration of osmosis and reverse osmosis. The arrows denote the water flow direction.

In RO the water will flow in the opposite direction compared to natural osmosis, Figure 5.1. This will require that a pressure is created.

The salty water is given a high pressure (Δp) that exceeds the osmotic pressure ($\Delta \pi$). Then the water from the concentrated solution will pass through a semi-permeable membrane and leave the solid salt particles behind. The osmotic pressure for seawater with 3.5% salt is 2.6 *MPa* (or $\approx$26 bar). Brackish water needs less energy since the osmotic pressure is lower. This means that a RO system for salty water only begins to produce water when a pressure higher than the osmotic pressure is achieved. In fact, the flux of the water through the membrane is proportional to the difference in pressure between the applied pump pressure Δp and the osmotic pressure $\Delta \pi$:

$$J_w = C \cdot (\Delta p - \Delta \pi)$$

where J_w is the flux (ℓ/m^2 membrane per hour) and C the so-called permeability constant. Consequently, the flow rate is

$$Q = J_w \cdot A = C \cdot (\Delta p - \Delta \pi) \cdot A$$

where A is the surface area of the membranes. The permeability depends on temperature. The higher the water temperature the higher the permeability will be. The change in permeability is about 3% per °C. As a result, the required pressure to achieve or keep a certain flux or capacity will be lower at higher temperatures.

A small part of the dissolved substance also goes through the membrane with the water. Some 2% of common salt (NaCl) may go through the filter in RO.

The actual required pressure and subsequent energetic cost of desalination is 2–3 times higher than the osmotic pressure due to inefficiency, material losses and membrane fouling. Typical operating pressure for seawater desalination is in the range *5.5–6.2 MPa* (or 55–62 bar) but pressures in the range *6.9–8.3 MPa* (69–83 bar) can be found.

Brackish water desalination needs less energy since the osmotic pressure is lower. It decreases almost linearly with decreasing salt content. The osmotic pressure (in bar) can be estimated with a rule of thumb formula:

$$\Delta \pi \approx 0.7 \cdot 10^{-3} \cdot C$$

where C is the salt concentration (*mg/ℓ*). For 1% salinity the osmotic pressure is around 0.75 *MPa* (or 7.5 bar).

> Energy for desalination decreases almost linearly with decreasing salt content.

Brackish water reverse osmosis (BWRO) systems typically require pressures in the range 1.5–2.5 *MPa* (or 15–25 bar). Consequently, less power is needed for brackish water desalination (BWRO) compared to seawater desalination (SWRO). Also, the lower pressures found in BWRO systems permit the use of low-cost plastic components. Therefore, the total cost of water from brackish water is considerably less than that from seawater, and systems are beginning to be offered commercially.

The required pressure-pump power is proportional to the flow rate of the feed and to the pressure difference between the exit and inlet pressure of the primary pump. The RO plants are sensitive to the feedwater quality, such as salinity, turbidity and temperature. Energy is consumed not only to achieve a sufficiently high pressure for the RO process; it is also needed for the pumping that pulls the water into the desalination facility as well as for the filters for the pre-treatment. Up to 72% of the total energy required for seawater desalination with RO membranes is consumed by the high-pressure pumps (Voutchkov, 2012). Therefore, it is essential to make sure that the pressure pump has a high efficiency.

5.3.4 Reverse osmosis membranes

In an RO membrane the water moves by flowing through the polymer structure, which is always hydrophilic and "swelled" by the water. The RO membranes are arranged in spiral-wound modules so that the membrane unit looks like a tube.

It is very important to pre-treat the feedwater. Suspended particles need to be removed via pre-filters and chemicals may be added ahead of the RO filters. A well-functioning pre-treatment is the key to reliable fresh water production. This will minimise membrane washes and will give the RO modules a longer life. Still membranes can be affected by fouling that can be caused by high-turbidity feedwater where membranes are clogged with suspended solids, marine organisms and their metabolic products. Fouling is also a result of the sedimentation of natural organic material (NOM) and mineral particles on the membrane surface.

Fouling restricts water flow and ultimately affects the water recovery of the membrane system. It can also damage membranes. Fouling can be prevented by adding special chemical agents or anti-scalants. The presence of fouling increases energy use. Frequent cleanings and chemical injection reduce the RO membrane life.

Even with good pre-treatment design it is vital to provide for periodic washing of the modules. If washing is inadequate, it generally must be repeated quickly.

The desalination process produces two streams of water, one with the fresh water (having low salinity) and one with more salt. From the feedwater flow about half the volume is converted to drinkable water and the other half will be about twice as salty as the incoming water. The second stream is called brine, and its disposal is a key issue. The brine is denser than the seawater. As noted before, chemicals like anti-scaling agents and coagulants are added to the seawater in the desalination process. The chemicals do not pass through the membrane and are left in the brine.

As soon as the brine is released, biological activities will begin, and microorganisms will feed on the chemicals. This consumes dissolved oxygen in the receiving water. Phosphate and nitrate are also released when chemicals in the reject water break down, causing eutrophication of the receiving water. So, if the brine is discharged into the sea, there is a risk for marine life. Too much salt can be just as deadly for sea life as seawater is for land animals and crops. Furthermore, the brine is often quite hot, so its disposal can have a negative impact both on land and in water. One way to minimise the potential negative effects of the salts and the chemicals in the brine is to mix the reject water rapidly with the surrounding water.

A desalination plant located at the ocean has practically unlimited access to seawater. Inland desalination plants, on the other hand, can be used for water reuse and drinking water production from used water where traditional sources are inadequate. For inland plants the disposal options must be carefully considered. The primary environmental concern with the disposal of concentrate to surface water, to sewers or by land application, is salt-loading the receiving waters, whether they be surface water or groundwater.

Sometimes it is possible to collect the brine in an evaporation pond. In a hot climate the water in the brine evaporates and the salt can be collected and used for other purposes.

5.3.5 Renewable energy for desalination

The energy required for desalination has three functions: to perform the pre-treatment, supply the high-pressure pump for the RO and overcome the membrane's resistance to the flow of the water. The energy is electrical and can come from a wide range of sources. Desalination based on the use of renewable energy sources can provide a sustainable way to produce fresh water. Currently an estimated 1% of desalinated water comes from energy from renewable sources, mainly in small-scale facilities. But larger plants are starting to add renewables to their energy portfolio.

Typically, the energy requirement for RO of seawater is 3–5 kWh/m^3 (IEA-ETSAP & IRENA, 2013) for the whole process. Some sources say that 1.5–2 kWh/m^3 is achievable to desalinate seawater. The theoretical limit for RO is around 1 kWh/m^3, while the practical limit seems to be around 1.5. For brackish water the energy required is in the range 0.5–2.5 kWh/m^3.

Solar PV and wind power can provide necessary energy, but electric energy storage is an important challenge, considering the intermittent nature of the production (Chapter 10). An important aspect is that excess energy can be stored as produced desalinated water, which can be a cost-effective storage solution when generation exceeds demand.

Excess energy can be stored as produced desalinated water. This can be a cost-effective storage when generation exceeds demand.

Thermal desalination requires both electricity and thermal energy, and – in total – more energy than the membrane process. Seawater desalination via MSF consumes typically 80 kWh/m^3 of heat energy (or 290 mJ/m^3) plus 2.5–3.5 kWh_e/m^3 of electricity (IEA-ETSAP & IRENA, 2013).

Bearing in mind the remarkable cost reduction of renewable technologies, desalination via renewables can already compete with conventional systems in remote regions where the cost of energy transmission and distribution is higher than the cost of off-grid generation. Desalination based on renewable energy is mostly based on the RO process, followed by the thermal MSF process. In existing applications solar PV is the energy source for almost half of the installed capacity, followed by solar thermal and wind. The best combination of renewable energy source and desalination technology will depend on the geographical location, the salinity of the water, the available renewable energy source and plant size. Of course, the requirements to pump and to pre-treat the feedwater are other factors to consider.

The cost for the solar cells is still the largest part of the cost, but batteries for electricity storage should also be considered (see Chapter 10). The batteries need maintenance. Any advances in storage capacity will have a profound impact on solar PV operations. On the other hand, the "fuel" is free, so the renewables will pay off with time, in contrast to diesel or grid electric power.

Solar energy is the most readily applicable source of renewable energy to be integrated with desalination technology. It can produce the heat and electricity required by all desalination processes. Worldwide, various renewable energy sources are used for desalination, as illustrated in Figure 5.2.

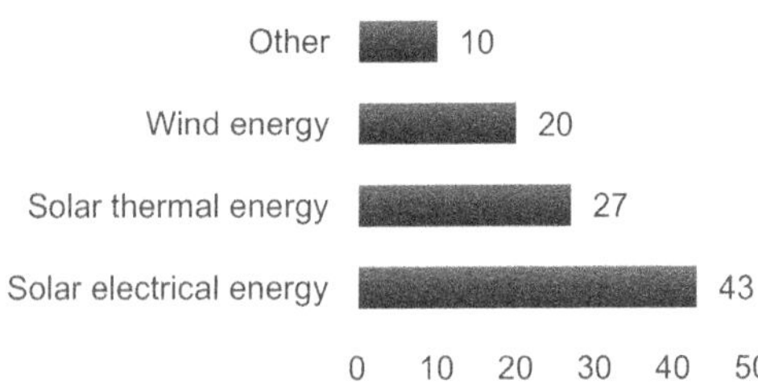

Figure 5.2 Renewable energy sources for desalination worldwide (in %). Data from Alkaisi *et al.* (2017).

Solar PV powered reverse osmosis has been proven commercially viable with significant reduction in production cost. One example is from Abu Dhabi. The levelised cost of large-scale desalination

(>100,000 m^3/day) is 0.91 USD/m^3 (3.3 Emirati dirham AED/m^3), compared to the current average water production cost of 1.42 USD/m^3 (5.16 AED/m^3) (Masdar, 2018).

Energy is the largest variable cost for seawater RO (SWRO) plants, varying from a third to more than half of the cost of produced water. According to Voutchkov (2016) there are no major technology breakthroughs expected in the next few years to dramatically lower the cost of seawater desalination. However, there is a steady reduction of production costs and technology advances. As Table 5.3 demonstrates, the cost will decrease significantly in the next few years.

Table 5.3 Reverse osmosis costs for medium and large best-in-class projects (Voutchkov, 2016).

	Year 2016	By 2021	By 2035
Cost of water USD/m^3	0.8–1.2	0.6–1.0	0.3–0.5
Electric energy use kWh/m^3	3.5–4.0	2.8–3.2	2.1–2.4

As mentioned, the source of energy, the water source salinity, the plant size and cost of land are factors that will affect the desalination cost. Typical desalination costs can vary from 0.5 to 3 USD/m^3.

***Example 5.1:** Required Power for Small-Scale Desalination*

Assume that we wish to produce 1 m^3 of desalinated seawater per day. To be conservative, this will require 8 kWh using RO technology. Assuming that a solar PV-based array can deliver power for six hours, this will require an electric power of 1.3–1.4 kW. In a subtropical or tropical country, a solar array of 2 kW_p would be sufficient to provide this average power during the day.

If the plant were downsized to 0.1 m^3 per day the required power would be 0.2 kW_p.

5.3.6 Operation and maintenance issues

There were some important lessons learnt from early installations of RO desalination powered by renewable energy (Cipollina *et al.*, 2015). Solar PV will only produce electric power in daylight. This has consequences both for the design of the renewable energy source and for the operational life of the membranes (see also Chapter 10).

The main maintenance issues are the cleaning of the solar panels and the difficulty of replacing RO pumps and modules. The most essential operational problems reported are:

- Fouling in membrane modules that requires cleaning or replacement;
- The booster pumps in the RO system (producing the high pressure).

The most common failures reported are electronic device breakdowns and sensor failures.

5.4 DISINFECTION

Disinfection is a key operation in any water supply. Here we emphasise the potential of using UV light as a potentially important and realistic opportunity to meet the challenge of pathogens in the water.

5.4.1 Disinfection technology

Eliminating harmful organisms in the water is essential to protect people's health. In water disinfection pathogenic microorganisms are deactivated or killed. Pathogens cause waterborne diseases such as cholera, polio, typhoid, hepatitis and several other bacterial, viral and parasitic diseases. Naturally it is of the utmost importance to substantially reduce the total number of viable microorganisms in the water.

Disinfection can be achieved by using either physical or chemical disinfectants. The most common physical disinfectants are ultraviolet (UV) light, electronic radiation and heat. Common chemical agents for disinfection are chlorine (Cl_2), chlorine dioxide (ClO_2) and ozone (O_3). In municipal water supply systems chemical inactivation of microbial contamination in the raw water supply is commonly used as one of the last steps to reduce pathogens in drinking water. Traditionally, chlorination is widely used for disinfecting water supplies in most western countries. For small water flow rates (household to village size) the primary treatment methods are ozone, UV radiation and chlorine. Techniques such as filtration may remove infectious organisms from water. However, filtering is no substitute for disinfection.

5.4.2 UV light disinfection

One major advantage of using UV light in remote areas is that it does not require any consumable chemicals. Maintenance is straightforward and there is no risk of overdosing. UV radiation does not leave any residuals in the water. UV light has been used quite extensively for water supply disinfection in small communities. It is one of the few affordable technologies for small-scale water supply that effectively kills most bacteria, viruses and other harmful microorganisms. A UV lamp will imitate sunlight. In nature sunlight will destroy some bacteria, purifying water naturally.

The efficiency of UV disinfection depends on the intensity and the wavelength of the radiation. If the water contains colour or turbidity, then the exposure of the microorganisms will decrease, and the disinfection becomes less efficient. This is of course a disadvantage. Therefore, some pre-filtering before UV radiation may be needed. Another problem is that there is no simple test of the disinfection result.

Disinfection with UV is usually done so that the water is passing though transparent pipes. It is sufficient to have a contact time of a few seconds. A UV dose (energy) is normally expressed in $mJ/cm^2 = mWs/cm^2$, the product of the UV intensity in mW/cm^2 and the contact time. A common dose is 20–40 mWs/cm^2 (EPA, 2011) to inactivate most waterborne pathogenic bacteria. The Department of Health and Human Services (U.S.) has established a minimum exposure of 16 mWs/cm^2 for UV disinfection systems. Most manufacturers provide a lamp intensity of 30–50 mWs/cm^2. In general, coliform bacteria, for example, are destroyed at 7 mWs/cm^2 (Oram, 2014).

The usual wavelength is 254 nm. A typical low-pressure UV lamp has a power rating of 40–85 W and will last for about 12,000 hours (15 months). It has an operating temperature of about 40°C (*ibid.*).

Typical power requirement to disinfect water with UV light is 10–20 $W/m^3 \cdot h$. For example, assume that we need to disinfect 1 m^3 of water, which is produced during six hours of sunlight. With a flow rate of 1/6 m^3/hour this will require a UV light power source of 1.7–3.4 W, a very small amount of power compared to the requirement for desalination.

The cost of a UV disinfection system is much lower than ozonation and membrane filtering. For ozonation the capital cost is roughly five times higher and the operating costs three times higher. Membrane filtration has a capital cost around ten times higher and an operating cost eight times higher (*ibid.*). Using RO technology is the best method to purify the water, since these membranes can remove all particles down to small molecules, but it comes at a price.

5.5 FURTHER READING

There is a lot of literature on desalination. For the non-specialist there is a good description of desalination and disinfection in Varadi *et al.* (2018) Chapter 5.3. Burn and Gray (2014) present a comprehensive description and analysis of reverse osmosis.

The American Membrane Technology Association (AMTA) has produced several easy-to-read fact sheets concerning RO membranes, filtration and desalination, including planning, operation and maintenance (AMTA, 2018).

The book Liehr *et al.* (2018) documents an ambitious decade-long project in Namibia. The book records experiences of rainwater and floodwater harvesting, groundwater desalination, sanitation and water reuse. Not only technology solutions but also social aspects, management and governance issues, economic viability and sustainability evaluation are described.

Bazargan (2018) presents a broad introduction to and overview of desalination and covers both technical and non-technical issues.

Drioli *et al.* (2011) describe the basics of RO and desalination. Bundschuh and Hoinkis (2012) discuss the applicability of renewable energy for fresh-water production, the various barriers and how to overcome them. Mahmoudi *et al.* (2017) present a comprehensive report on renewable energy and desalination. Shatat *et al.* (2013) review the global opportunities for solar desalination.

Chapter 6

Solar thermal desalination and solar water heating

> "Solar power is the last energy resource that
> isn't owned yet – nobody taxes the sun yet."

Bonnie Raitt, American blues singer and songwriter.

> "It's time for the human race to enter the solar system."

Dan Quayle, Vice President of the United States 1989–1993.

It is not necessary always to use electric power to produce clean water. Solar distillation is described in 6.1 as a complement to other water supply methods. Heating water has been done with solar thermal collectors for a long time. In 6.2 it is shown to be a complement to using electric power for heating and cooking.

6.1 SOLAR STILL DISTILLATION FOR CLEANING WATER

Solar still distillation (SSD) is simply a natural evaporation-condensation process. This is a practical and simple technique to clean water in remote arid areas and does not require any electric power, only the heat from the sun. The advantages of an SSD system include

© IWA Publishing 2018. Clean Water Using Solar and Wind: Outside the Power Grid
Author: Gustaf Olsson
doi: 10.2166/9781780409443_0079

low investment costs, low maintenance and low energy requirements. It is certainly environmentally friendly. However, an SSD system has low productivity; but for small-scale water demands, those of a small village or a household, SSD is a viable option due to the high cost of water transportation. It also offers a complement to electric power-driven water supply so that the available electric power can be used for other purposes.

The key function of an SSD system can be illustrated by a typical greenhouse. A constant volume of brackish or saline water is enclosed in a basin with a dark bottom. The basin can be constructed of concrete or some fibre-reinforced plastic. The roof is a transparent material like glass or plastic. The brackish or saline water is fed to the basin. The distillate is collected at the lower end of the roof. As the sunrays pass through the glass roof they are absorbed by the blackened bottom and will heat the water. The vapour pressure will increase and the water vapour is condensed on the underside of the roof. It will run down into troughs that conduct the distilled water to a storage basin. The process is a direct application of the greenhouse effect. Furthermore, the roof encloses all the vapour, prevents its loss and keeps the wind from reaching and cooling the contaminated water.

As the saline water is distilled there will be increasingly higher salt concentrations in the basin. Therefore, the systems need flushing, which preferably should be done during the night. There are several designs of SSD systems and some of the more popular designs are described and reviewed by Kalogirou (2005) and Al-Karaghouli and Kazmerski (2011). A typical still production is about 3–4 l/m^2 per day (Daniels, 1977), but 8–10 l/m^2 per day has been reported in simple and practical designs.

The real advantage of the SSD system is its simplicity and low cost. Since it is less efficient it needs much more space than the desalination methods described in 5.3 to produce a given volume of water.

6.2 SOLAR WATER HEATING

Solar thermal collectors or solar water heating (SWH) have been used for a long time as heating sources for water. The technology is simple and extensively proven. For cooking this may remain the backbone of water heating, even if electrification of the heating process is possible. Solar thermal collectors can be installed on roofs and work well in climate zones all over the world. In sunny regions in Africa and Asia they will naturally be an important complement to solar PV systems.

Active solar heating systems use a *collector* and a fluid that absorbs solar radiation. Fans or pumps circulate air or heat-absorbing liquids through collectors and then transfer the heated fluid directly to a room or to a heat storage system. Active water heating systems usually have a tank for storing solar heated water.

Solar systems for heating water or air usually have non-concentrating collectors. This means that the collector area (the area that intercepts the solar radiation) is the same as the absorber area (the area absorbing the radiation). *Flat-plate collectors* are the most common type of non-concentrating collectors and can produce water with a temperature of more than 90°C.

The surface of the absorber determines the efficiency of solar water heaters. It has to combine high absorbance (percentage of incoming energy that a material can absorb) and low emittance (percentage of energy that a material radiates away) of the solar radiation. At the top and bottom of the metal absorbing plate, thicker copper pipes, known as headers, assist in the removal of heated water and the arrival of colder water to be heated.

The principles of operation for flat-plate collectors are fairly consistent, but significant improvements in the design of systems, particularly absorber plates, have occurred. The major types of thermal collectors are:

- *Evacuated tube solar thermal system*: this is one of the most popular solar thermal systems in operation. The collector itself is made up of rows of insulated glass tubes that contain copper pipes at their core. Water is heated in the collector and is then sent through the pipes to the water tank. This type of collector is the most efficient, but also the most expensive.
- *Flat-plate solar thermal system*: this is a type of device that has been in use since the 1950s. The main components of a flat-plate panel are a dark coloured flat-plate absorber with an insulated cover, a heat-transferring liquid to transfer heat from the absorber to the water tank and an insulated backing. The flat-plate feature

of the solar panel increases the surface area for heat absorption. The heat-transfer liquid is circulated through copper or silicon tubes contained within the flat surface plate. The flat-plate design is slightly less compact and less efficient when compared with an evacuated tube system, but the cost is lower. A flat-plate SWH can work well in all climates and can have a life expectancy of over 25 years.

An SWH system contains a storage tank. In a close-coupled system the storage tank is mounted with the collector on the roof. The tank is located above the collectors to take advantage of the principle of thermo-siphoning. This principle utilises collectors to heat the water, which rises to the storage tank. Cold, denser water flows through the collector, is heated up and is then returned to the tank. As the heated water is less dense, it rises to the top of the tank. Roof-mounted flat-plate collectors that utilise thermo-siphoning are extremely popular.

Since the system has no moving parts it is easy to maintain and reliability is very high. How about electricity requirement? Yes, electricity may be needed to (1) lift the cold water to a level where it can flow into the storage tank and (2) make sure that the system does not overheat. If the storage tank is filled with hot water, it may overheat and a valve may be closed so that no more hot water is fed into the tank until more cold water is fed into the system.

6.3 FURTHER READING

Several practical SSD solutions are demonstrated on www.youtube.com (search for "solar: still distillation"). Kalogirou (2004) presents a survey of the various types of solar thermal collectors and applications.

Practical demonstrations as well as academic lectures on solar thermal systems can also be found on www.youtube.com (search for "solar thermal").

Chapter 7

Used water treatment

"Clean water is a great example of something
that depends on energy. And if you solve the
water problem, you solve the food problem."

Richard Smalley, American scientist, 1943–2005.

Clean air and clean water are absolute top priorities when we talk about responsible energy development. Over the last few decades there has been an interesting development in wastewater (or used water) treatment. The traditional method of used water treatment has been centralised sewage treatment systems connected to urban water users by wide sewer networks. Now there is a remarkable development into decentralisation, and it has been recognised that centralised systems have their limitations. This becomes obvious looking at the need for water remediation in remote areas. Small-scale treatment must be both efficient and affordable. Water scarcity makes it necessary to reconsider the whole concept of sewer-based water treatment. Every drop of water is important.

Sanitation is still out of reach for too many people, as mentioned in Chapters 1–3. Furthermore, both clean water and sanitation must be achieved in order to reach the UN's Sustainable Development Goal

© IWA Publishing 2018. Clean Water Using Solar and Wind: Outside the Power Grid
Author: Gustaf Olsson
doi: 10.2166/9781780409443_0083

number 6. Having available energy is just one of the key requirements. If individual households can be empowered, many of the sanitation problems in remote areas can be solved locally without the support of governments or other institutions.

The main philosophy that should guide the wastewater treatment is that wastewater is not waste: it is a renewable and recoverable source of energy, nutrient resources and water. This should guide the development of used water treatment, considering:

- Recovery and utilisation of energy content present in the water;
- Recovery of nutrients;
- Production of water for reuse.

All three aspects can be satisfied with proven technology combined with renewable energy of realistic size. Decentralised used water treatment is described in 7.1. Various technologies are briefly explained in 7.2 and energy aspects to treat the water are summarised in 7.3.

7.1 MAIN SOURCES OF USED WATER

Decentralised treatment plants are based on self-contained modular systems that provide efficient, scalable solutions to used water treatment problems of all sorts. To find out suitable technologies for decentralised water recovery treatment it is necessary to identify the various sources of water contamination. The traditional method of used water treatment has been to treat a single waste stream from a sewer system. A decentralised system can be designed in a different way with a separate treatment facility for each category of contaminated water. In this way we may look at urine, faeces and greywater as separate sources of used water instead of being components of one type of used water. The sources should be considered useful resources rather than waste. This will also help when possibilities for financing the treatment are considered (see Chapter 12).

Depending on the source the used water can be treated differently. Here we consider three major sources from households:

- Urine,
- Greywater from washing, and
- Blackwater from toilets.

Urine is an important source of sustainable nutrients and is usually less contaminated than water containing faeces. It is a major contributor to nitrogen, phosphorus and potassium and can be an excellent fertiliser. Urine must be separated from faeces, which can be used for biogas production. Urine can be stored in a tank until it is needed for agriculture; and if it is stored for a sufficiently long time it becomes hygienised and safer for agricultural use (WHO, 2006). Storage space may be available in remote areas but more difficult to come by in urban or peri-urban areas. A storage tank will need pumping capacity and the maintenance demand should not be underestimated (Tilley, 2013).

Greywater is often a source of many pollutants, an overview of which is presented by Friedler *et al.* (2013). High concentrations of organic substances come from the kitchen sink. Typical components are food residues, oils and fats, detergents and drain cleaners. Greywater from washing usually contains shampoos, soaps, preservatives and dyes. The water may also contain some heavy metals like zinc and copper from in-house plumbing.

Faeces and blackwater contain high concentrations of organic matter and are useful sources of bioenergy. Anaerobic treatment is a method of extracting the inherent energy and producing biogas. Particular attention should, however, be paid to hygiene; pathogens must be reduced or eliminated.

Naturally, separating the used water into different categories for separate treatment increases the complexity of the process but it is an important way to save precious water resources.

7.2 TREATMENT OF USED WATER

Today there is a wide spectrum of methods and processes available for small-scale treatment of used water. On top of conventional methods there are treatment technologies developed for specific products. However, not only do technologies have to be implemented, but financing and business models should be considered as well, as pointed out by Tilley (2013). It should be emphasised that no particular technology is necessarily better than another. The key point is the relevance of the technology for the local conditions and how its users are motivated to manage it. Here we describe major technologies that are sufficiently simple in design and operation and possibly affordable for low-income people. Primarily, we consider the energy aspect.

7.2.1 Septic tanks

A septic tank consists of a concrete or plastic tank. The design of the tank usually includes two chambers separated by a dividing wall with openings located about midway between the floor and roof of the tank. Used water enters the first chamber where solids will settle. These are anaerobically digested (see below), reducing the volume of solids. The liquid flows through the dividing wall into the second chamber, where further settlement takes place. The excess liquid, now in a relatively clear condition, then drains from the outlet into a septic drain field.

The remaining impurities are trapped and eliminated in the soil, with the excess water eliminated through percolation into the soil, through evaporation and by uptake through the root systems of plants and eventual transpiration, or entering the groundwater or surface water. The required size of the drain field must increase with the volume of used water. On the other hand, if the soil has a higher porosity, the drain field can be smaller compared to a field with lower porosity. The entire septic system can operate by gravity alone. A lift pump will be required for certain places.

Obviously, a septic tank has two major drawbacks: the inherent energy in the organic matter is not used and the effluent water is not reused.

The organic content of the waste is converted to biogas (mainly methane) during anaerobic digestion. The gas will escape into the air, unnecessarily contributing to the greenhouse effect in consideration of the fact that the Global Warming Potential (GWP) of methane is 30–90 times higher than that of CO_2 depending on the considered timescale.

The other drawback is that the effluent water is not reused. Naturally, a septic tank can be considered the first step in a water reuse scheme, but the loss of organic energy is still there. Furthermore, the septic tank must be emptied from the settled sludge, typically every three to five years.

7.2.2 Activated sludge systems

The traditional process for removing organic matter is biological oxidation, which involves microorganisms feeding on the carbon and the oxygen in the water. Around half of the organic matter is used for the growth of the microorganisms, in other words to increase their body mass. The other half is converted into carbon dioxide.

Aeration to provide oxygen to microorganisms requires electric energy for a compressor that provides compressed air. This is normally the major part of energy demand for conventional treatment. There are many different treatment schemes using aerobic biological processes. The air is dissolved in the water, supplying the organisms with the necessary amount of oxygen for their metabolism. The oxygen supply has to be sufficiently high to satisfy the microorganisms. However, an excess of aeration will not favour the biological activity; it will only waste aeration energy. So, aeration should find a balance between the biological requirement of oxygen and the need to save energy.

The energy requirement for aeration motivates to replace aerobic (using air) treatment with anaerobic treatment (requiring absence of air, see below). The key motivation to use aerobic treatment is the rate of the biological reaction. In an aerobic treatment the speed is an order of magnitude higher. As a consequence, the plant volume can be smaller, but it implies an energy cost. Therefore, when space is available there is a rationale for replacing aerobic treatment with anaerobic treatment. Even more important: in anaerobic treatment the energy in the organic matter will be converted to biogas that becomes an important energy source.

In conventional systems the influent water contains not only organic matter but also nitrogen that principally arrives at the plant as ammonium NH_4^+ (60–80%). Most nitrogen removal plants will transform the ammonium into free nitrogen that will escape via the water surface. The removal of nitrogen is a slower process than the removal of organic carbon and takes place in two principal stages, nitrification and denitrification. In nitrification ammonium is transformed into nitrate NO_3^- (an oxidation process) and in the denitrification (a reduction process, where no oxygen is allowed) nitrate is reduced to nitrogen gas N_2.

The concentration of dissolved oxygen (DO) governs carbon removal, nitrification and denitrification. In carbon removal and nitrification, the process rate will increase with the oxygen concentration. However, there is a limit to the process rate, and higher DO concentrations will not help the biology but only waste energy for the compressors that aerate the biological reactor. With too little DO microorganisms will suffocate and the process rate will be significantly reduced. In extreme cases the organisms will die. The opposite applies to denitrification: the higher the level of dissolved oxygen, the lower the rate.

Aeration typically represents 50% to 60% of a sewage treatment works' energy demand but energy consumption as high as 75% has been reported. This fact explains why aeration control is key to saving energy (Olsson, 2015).

Since urine accounts for most of the nitrogen in used water, urine separation eliminates the need for nitrification and denitrification, as noted in the previous section. It also lowers the energy demand for the organic matter.

Obviously, there are more energy-efficient methods to use in remote areas than the activated sludge process. Electric energy should be used for purposes other than oxidising organic matter. Instead, the inherent energy in organic carbon should be converted to useful energy via anaerobic digestion to produce biogas.

7.2.3 Anaerobic digestion

Simple home- and farm-based anaerobic digestion systems offer the potential for cheap, low-cost energy for cooking and lighting by producing biogas. The organic material that provides the feedstock for the anaerobic digestion may consist of faeces and blackwater, manure from pigs and cows, food residues and agricultural residues. The production of biogas from anaerobic treatment is a key component in resource recovery. The biogas can be converted into energy, for example to provide heat for cooking.

Anaerobic bacteria are some of the oldest forms of life on earth. The same types of anaerobic bacteria that produce natural gas in nature also produce methane in the technical processes. Anaerobic bacteria evolved before the photosynthesis of green plants released large quantities of oxygen into the atmosphere. Anaerobic bacteria break down or "digest" organic material in the absence of oxygen and produce biogas as a waste product. Anaerobic decomposition occurs naturally in swamps, waterlogged soils and rice fields, deep bodies of water and in the digestive systems of termites and large animals.

In the anaerobic process microorganisms assist the process of organic material conversion that produces the biogas. The processes involved in fermentation are exceedingly complex and are not completely understood. There is an impressive amount of activity going on in anaerobic process research to increase knowledge of the microbiology and the internal mechanisms taking place.

A variety of factors affects the rate of digestion and biogas production. The most important is temperature. Anaerobic bacteria communities can endure temperatures from below freezing to more than 60°C and different species of bacteria thrive in different temperature ranges. At temperatures in the range 35–40°C there are mesophilic bacteria. Some of the bacteria, thermophiles, can survive at temperatures around 50–65°C. Mesophile bacteria are generally more tolerant than thermophiles of changes in environmental conditions. Therefore, mesophilic digestion systems are considered to be more stable than thermophilic ones. However, at increased temperatures the reaction rates are faster and consequently the gas yield is higher. The hydraulic retention time is also shorter. Therefore, smaller reactor volumes are required under thermophilic conditions. At lower temperatures, from 35°C to 0°C, the bacterial activity, and thus biogas production, falls off gradually. Rapid changes of the reactor temperature will upset the bacterial activity; therefore, the digester must be kept at a consistent temperature. The digesters referred to here are in subtropical or tropical regions, so they seldom need extra heating to enable their operation.

Anaerobic digestion (AD) has more advantages than the energy generation via biogas: it has a high capacity to treat slowly degradable substrates at high concentrations and it can efficiently reduce pathogens.

7.2.4 Membrane separation

Membrane separation has been described in Chapters 5.2 and 5.3. With the whole spectrum of membrane processes, it is possible to remove micron-sized particles like microorganisms and suspended solids. A partial or full disinfection can also be achieved. As mentioned in Chapter 5, the membrane technology is compact and provides consistent quality over a range of pollutant loading rates. It depends on energy supply. The main challenge is to maintain a consistent throughput capacity even if the product quality is unfailing. Thus, membrane fouling and cleaning are of major importance. A good overview of membrane technology in used water treatment is found in Leslie and Bradford-Hartke (2013).

7.2.5 Disinfection

Disinfection has been described in Chapter 5.4. The aim of any type of sanitation system must be to inactivate or reduce potentially pathogenic organisms. Since a decentralised system often aims at water reuse, for example when greywater is used for personal hygiene, it becomes crucial to reduce the pathogens. As mentioned in Chapter 7.1, there are international guidelines for hygiene in water (WHO, 2006). Stenström (2013) has described major issues in reducing pathogens in water. Storage of urine has been mentioned as an effective way of reducing pathogens. In faecal excreta, on the other hand, a much longer time is needed to reduce the risk of pathogens. The hygienisation depends on the storage time, pH and temperature. The storage time is usually more than a year. Thus, anaerobic treatment is advantageous, both to produce energy and to reduce the pathogens.

7.3 ENERGY ASPECTS

The main operations of used water operations that depend on electric energy are:

- Pumping;
- Aeration;
- Disinfection.

Energy for pumping has been discussed in Chapter 4. Three aspects are essential: the required flow rate, the elevation of the water and the timing of the pumping. The various types of used water may be stored in tanks and can be brought into the tanks mostly by gravity. The timing of when the tanks are to be emptied is not critical, which is helpful when electricity generation is intermittent.

For activated sludge plants the typical energy use in Sweden, Germany and Switzerland is of the order 15–45 *kWh*/person/year, where small plants have the higher specific-energy demand. Aeration is the dominating user or electrical energy (Lingsten *et al.*, 2011; Balmér, 2010). Assuming the high-end western European energy use of 45 *kWh*, each person would need an average power of 26 W over the year, similar to low-energy light bulbs. As an example, a small activated sludge plant for one household has been in operation in Sweden over the last year (Gillblad, 2018). A power supply of 60 *W* has been more than adequate, since most of this power is used for an UV lamp for disinfection. Compared with other uses the aeration energy requirement is small. However, aeration requires a continuous supply of electric power. This must be considered for solar and wind power supplies, so energy storage is important.

Anaerobic treatment needs very little electric energy, mostly for some mixing of the sludge in the digester. This is an almost negligible energy use compared to other loads.

Membrane separation needs electric power for the compressor delivering the pressure to the membrane process. As noted in Chapter 5.3, the required pressure depends on the pore size of the membranes. For micro- and ultra-filtration, the required pressure is much less than for reverse osmosis, so the necessary electric power input is much lower than for desalination (see Chapter 5.2 and 5.3).

Disinfection using UV lamps is not an expensive energy user, as noted in Chapter 5.4.

7.4 FURTHER READING

There are hundreds of books and a huge literature on used water (or wastewater) treatment. This book does not aim to be a textbook on water reclamation, but here are a few standard references:

Grady *et al.* (2011) give a comprehensive basis for the understanding of biological wastewater treatment.

Metcalf and Eddy (2013) is a widely known standard textbook on wastewater treatment design and Asano *et al.* (2007) is devoted to various methods for water reuse.

Energy use in wastewater treatment is examined in the book Olsson (2015).

Part III

Renewable Energy Technologies

Renewable energy is revolutionising energy availability. Climate change motivates the search for energy sources that do not generate a carbon footprint. It must be recognised that traditional energy sources require water for primary fuel extraction, refining, transportation and electricity generation. Solar PV and wind energy can decouple the dependency between water and energy with much smaller carbon and water footprints. The other crucial property is the scalability of solar PV and wind, from stand-alone household production to villages and regions. However, the sun does not shine all the time and the wind is sometimes too calm. Therefore, the intermittent generation of energy from solar and wind has to be managed. Stored energy must be available when the primary sources are not available.

Solar PV is the topic of Chapter 8 and wind energy is briefly discussed in Chapter 9. Various ways to handle the intermittent nature of these energy sources are described in Chapter 10. Some issues of energy management are discussed in Chapter 11. The coordination of different production units and different loads is critical. The use of energy storage capacity makes it necessary to carefully plan the energy flow in the stand-alone system.

Chapter 8

Solar PV

> "Solar power is a safe form of nuclear energy. We are using fusion
> reactions that are 150 million *km* away to make light that we
> then convert to electricity with photovoltaic modules."

Professor Sean White (2016).

The end user of a solar photovoltaic (PV) system does not need to know
any details of the technology to convert sunlight to electric power. It
is sufficient to know that a certain number of solar cells will produce
a certain amount of electric power. However, it may be useful to have
some background knowledge about the solar technology to estimate
what can be expected in terms of both productivity and potential
problems.

In Chapter 8.1 the fundamental properties of sunlight and its
influence on solar panels are described. Characteristic parameters of
solar cells are explained in 8.2. Solar cell technology – how sunlight is
converted to electricity – is briefly explained in 8.3. It is important to
realise the importance of solar cell efficiency. Solar cell temperature is
a key parameter determining efficiency. Solar cells are combined into
systems, as described in 8.4. The solar PV energy potential for various

© IWA Publishing 2018. Clean Water Using Solar and Wind: Outside the Power Grid
Author: Gustaf Olsson
doi: 10.2166/9781780409443_0095

water operations is illustrated in 8.5. Economic aspects of solar PV operations are further reviewed in Chapter 12.

8.1 UTILISING THE SUN

The solar radiation that reaches the outer atmospheric layers of the Earth has a particular spectral profile. Its intensity is highest between the wavelengths 0.3 and 4 μm. About 7% of the extra-terrestrial spectrum energy intensity is at ultraviolet wavelengths, 47% in the visible range and 46% infrared.

8.1.1 Irradiance

Irradiance is the measure of solar *power* at a certain geographical location and is measured in W/m^2 (see also 3.5). A significant amount of solar radiation is attenuated as it travels through the atmosphere. This attenuation is due to:

- Absorption of solar radiation by different particles in the atmosphere,
- Backward scattering and reflection of solar radiation by air particles, water vapour, dust etc.

On a *clear* day some 80% of the radiation reaches the ground and is absorbed. Around 5% is scattered and reflected into space. 10–15% can be absorbed by particles in the atmosphere and typically 2% is reflected from the ground.

On a *cloudy* day 30–60% of the radiation can be reflected and 5–20% absorbed by the clouds. There is a diffuse radiation from the clouds, which reaches the ground, so the total irradiation at ground level may be 0–50% of the radiance. A solar irradiance of 1,000 W/m^2 is known in the PV community as *one equivalent sun*.

8.1.2 Global horizontal irradiance

A solar cell will absorb a combination of several parts:

- Direct radiation comes as a direct beam from the sun to the solar cell;
- Ground-reflected radiation is the part of the radiation that is first reflected on the ground and then reaches the solar cell.

The irradiance is quantified as *Global Horizontal Irradiance* (GHI) and is the total amount of solar energy incident on a horizontal surface, coming from both direct and scattered sunlight. GHI is a key parameter for solar PV applications. The GHI varies quite a lot depending on the geographical location: see Table 8.1. For most of the off-grid locations that we are focusing on – rural areas in Africa and developing Asia – the GHI is high. There are several sources of information of the solar irradiance around the world, for example the Global Solar Atlas, published by the World Bank Group (http://globalsolaratlas.info).

Table 8.1 Annual Global Horizontal Irradiance (GHI) in various regions of the world.

Region	GHI Range kWh/m² Per Year
Africa	1,600–2,700+
Middle East & North Africa	1,700–2,700+
Latin America & Caribbean	1,000–2,700+
North America	<700–2,600+
Europe	<700–2,100
South and Central Asia	1,400–2,400
East Asia	1,000–2,300
South-East Asia & Pacific	900–2,600

Source: Data from WEC (2016, Chapter 8).

The irradiation depends on the elevation angle α of the sun. For low elevation angles the direct solar radiation needs to cross a larger section of the atmosphere to reach any point on the ground, so the attenuation factor is larger. The length of the solar radiation path through the atmosphere compared to the shortest path is indicated as air mass (*AM*). *AM1*, i.e. one air mass, indicates a vertical path from sea level to the outer atmospheric layer at standard barometric pressure with the sun directly overhead. The *AM* is defined to be zero outside the atmosphere. A simplified, and widely reported, relationship links *AM* to the solar elevation α: $AM = 1/\sin\alpha$. This equation does not hold for very low values of the solar elevation α. Conventionally, *AM* is given a value of 38 for $\alpha \approx 0$.

The most widely used solar PV systems are flat panels. The useful power amount of a solar beam striking a plane surface is its normal component with respect to the plane. The solar power absorbed by the surface depends on the beam's angle of incidence θ with the normal direction from the plane. The maximum power transfer (h_0) is reached when the incident beam is perpendicular to the surface: that is $\theta = 0$. For angles different from zero, the share of useful power collected on the plane (h) is $h(\theta) = h_0 \cdot \cos\theta$.

8.2 SOLAR PV CHARACTERISTIC PARAMETERS

Solar PV cells generate DC power in rough proportion to the total incident solar irradiation. The PV power output also depends on the PV cells' operating temperature. As the temperature increases the output power will decrease, as shown in 8.3.

Solar electricity systems are given a rating in kilowatts peak (kW_p). This is essentially the rate at which it generates energy at peak performance, for example at noon on a sunny day. Obviously, the true electricity generation rate will depend on the system's orientation, shading and how sunny the location is. Solar irradiation of 1 kW/m^2 is used to define standard testing conditions (STC). To give an order of magnitude: for each m^2 of commercial solar cells there is about 140–200 W of electric power produced. In other words: the solar cell will have an efficiency of the order 14–20% (see also 8.3.2).

When sunlight hits a solar cell, it will show a typical characteristic relation between output current and voltage. Four parameters characterise the solar cell:

- The short circuit current I_{SC},
- The open circuit voltage V_{OC},
- The peak power P_{max}, and
- The fill factor FF.

These parameters will determine the efficiency of the cell.

If the module were to be short-circuited then the current I_{SC} would flow through the circuit, at least in theory. The short circuit current represents the upper limit of the current that is generated by the solar cell. It is proportional to the irradiation and the cell area. Commercial silicon solar cells can deliver a high current. For a cell of $150 \cdot 150$

mm² it typically reaches almost nine amps under standard conditions of 1 *kW/m²*.

If the cell electrodes are disconnected, then there is no current flowing through the external circuit. The voltage under this condition is called the open circuit voltage (V_{OC}) and is the maximum voltage that a solar cell can deliver. Typically, the value for commercial solar cells is of the order 0.4–0.6 *V*. The relation between the short circuit current and the open circuit voltage is illustrated by Figure 8.1.

Solar cells can deliver between 35 and 200 *W/m²*.

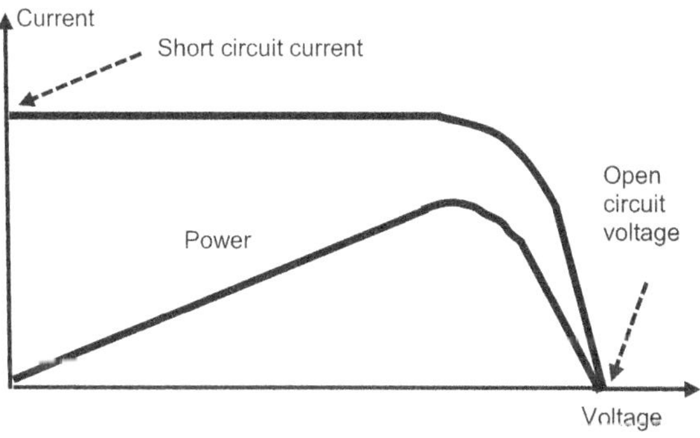

Figure 8.1 Principal relation between the current and voltage in a solar cell. The short circuit current I_{SC} is obtained by zero voltage and the open circuit voltage V_{OC} is reached by zero current (=open circuit). The maximum output power is reached by a certain combination of current and voltage.

The maximum power – called the peak power P_{max} (indicated as W_p or kW_p) – from a cell can be delivered at specific values of the output voltage and the corresponding current. This is illustrated in Figure 8.1. Naturally the electric circuitry should be designed so that the cell can deliver peak power under the given external conditions. Typically, a solar cell of 150 · 150 *mm²* can deliver a maximum power exceeding 5 *W* or of the order 200 *W/m²*.

The ratio between the theoretical maximum power generated by a solar cell (P_{max}) and the product of the open circuit voltage and the short circuit current is called the fill factor, FF:

$$FF = \frac{P_{max}}{V_{OC} \cdot I_{SC}}$$

The fill factor is typically around 0.75–0.8.

Based on the four characteristic parameters, the conversion efficiency can be calculated and is defined at the ratio between the maximum generated power P_{max} and the incident power (P_{in}) under standard testing conditions (STC):

$$\eta = \frac{P_{max}}{P_{in}} = \frac{V_{OC} \cdot I_{SC} \cdot FF}{P_{in}}$$

The efficiency is further examined in Chapter 8.3 and has typical values between 0.14 and 0.2.

8.3 CONVERSION OF SUNLIGHT TO ELECTRICITY

Solar PV technology uses a semiconductor material to directly convert sunlight to electricity. In such a material the absorption of the light raises an electron to a higher energy state. This electron will then move from the solar cell into an external circuit. The electron delivers its energy in the external circuit and then returns to the solar cell. Almost all PV energy conversion uses semiconductor materials, where a *p-n* junction has been created, to allow this to happen.

8.3.1 Photovoltaic technologies

There have been various photovoltaic technologies for solar cells developed to date:

- Crystalline silicon technology, either monocrystalline or polycrystalline,
- Thin-film, such as amorphous silicon, and
- Organic cells.

The dominating technology is based on either monocrystalline or polycrystalline silicon (c-Si) technology. Monocrystalline cells are produced by slicing wafers (up to 150 *mm* diameter and 200 microns thick) from a high-purity single crystal boule. Four sides of the silicon ingots are cut off to make high-purity silicon wafers. As a result, a large part of the original silicon becomes waste, which adds to the production cost.

Polycrystalline cells are manufactured by sawing a cast block of silicon first into bars, and then into wafers. The process is simpler than the process to make monocrystalline since less silicon waste is produced. The main trend in crystalline silicon cell manufacturing involves a move toward polycrystalline technology. There is an intense effort to reduce the costs to produce monocrystalline cells, so they might be more competitive with polycrystalline in the future. According to Fraunhofer (2016) silicon-based PV technology accounted for about 94% of the total production in 2016. The share of polycrystalline technology is now about 70% of total production.

Thin-film cells are constructed by depositing extremely thin layers of photosensitive materials onto a backing such as glass, stainless steel or plastic. Thin-film manufacturing processes result in lower production costs compared to crystalline technology. The most popular thin-film technology is amorphous (uncrystallised) silicon (a Si). In amorphous silicon solar cells there is much less silicon used (about 1%) compared to crystalline silicon cells. The cells can be grown in any shape or size and can be produced in an economical way. Amorphous silicon cells have been used for a long time in non-critical outdoor applications and consumer products such as watches and calculators. They are now being adopted by other larger-scale applications. In 2016, the market share of all thin-film technologies amounted to about 6% of total annual production (Fraunhofer, 2016). Mass production costs of thin-film modules are lower than for crystalline silicon modules. They also have a higher heat tolerance.

The drawback of the thin-film technology is its lower efficiency, due to the active material used. Commercially interesting materials are cadmium telluride (CdTe) and copper-indium/gallium-diselenide/ disulphide (CIS/CIGS).

There are several technologies based on organic cells. This includes dye-sensitised solar cells, antenna cells, molecular organic solar cells and completely polymeric devices. The dye cell is the closest to commercialisation; the other organic cells are still in their development phase.

In addition to thin-film and organic solar cells, compounds such as perovskite are being used in developing the next generation of solar PV cells. This is a mineral ($CaTiO_3$) that was discovered in the Ural mountains in Russia in 1839 and is named after the Russian mineralogist Lev Perovski (1792–1856). Efficiencies of perovskite cells have jumped from 3.8% in 2011 to over 20% by 2014. While perovskite solar cell technology is yet to be consolidated for commercial use, cells can be very cheap as they are made from relatively abundant elements such as ammonia, iodine and lead. However, there are severe doubts in terms of environmental safeguarding over the use of lead. Moreover, perovskite cells are unstable and deteriorate significantly when exposed to the environment. The replacement of lead with a more environmentally friendly element and the improvement of stability will be important objectives for researchers in this field. Developing a tandem solar cell with correct matching of the top perovskite and bottom silicon cell is expected to increase the conversion efficiency of the solar cells well beyond 25%, while keeping manufacturing costs low.

8.3.2 Efficiency of PV modules

A major concern about photovoltaic technologies is their efficiency, life and performance over time. Commercial PV modules convert around or less than 20% of the solar energy irradiation to electricity. The efficiencies of PV panels also have a temperature coefficient, and generally degrade in rising temperatures.

Efficiencies of solar PV modules are typically of the order 14–21%. Table 8.2 shows that monocrystalline panels have the highest efficiency. They also have the longest expected lifetime. Polycrystalline and thin-film solar modules have a lower cost but are less efficient. Therefore, there is a trade-off between production cost and efficiency to find the best choice of panel type.

> Efficiencies of solar PV modules are of the order 14–21%.

Table 8.2 Efficiencies of commercial solar PV modules.

Type of Panel	Efficiency (Commercial) %	Efficiency (Best Lab) %
Monocrystalline	17–21	24.4–26.7
Polycrystalline	14–16	19.9–22.3
Thin-film CIGS	14–16	19.2–21.7
Thin-film CdTe	14–16	18.6–21.0
Thin-film a-Si	13.8–15.5*	11–14

Source: Data from Fraunhofer (2015, 2016).
*Based on modules with highest efficiency of their class.

Research on solar cells is making an impressive improvement to available efficiencies. With a combination of concentrated solar PV and new semiconductor technologies (III–V multijunction concentrator solar cells) the achievable efficiency reached 46% in 2017 (Fraunhofer, 2016).

Solar panels have a relatively long lifetime and often have a 25-year performance guarantee. This is of course an unusually long time, so the solar panels are considered to be the most reliable parts of a solar system. There is still, however, a degradation in performance and power output decreases about 1% in the first ten years.

Solar panels have a commonly guaranteed lifetime of 25 years.

Another problem of efficiency loss is caused by what is called mismatch. One would assume that the cells in the sun would deliver their power to the load. Instead the shaded cells may absorb the power and significantly reduce the total efficiency of the module. If a single cell gets no light, then the current in all the series-connected cells will be reduced to the same current level as the shaded cell. So, one poorly performing cell (in the shade) can cause the total power output to be reduced to zero. More seriously, there may be a local heating of the module that will damage some cells. This local heating, however, can be avoided by using protective electronics – so-called bypass diodes – around the cell.

Energy is needed to produce solar cells. One way to express the energy requirement is to calculate how much time is needed for the solar PV system to produce as much energy as was required to manufacture the panels. The payback time naturally depends on the geographical location.

The payback time is displayed as function of the irradiation (in *kWh/m²/year*) in Fraunhofer (2016). The energy payback time (EPBT) for 1,000 *kWh/m²/year* (typically for northern Europe) is found to be 2.1 years, and for 1,700 *kWh/m²/year* (typically for southern Europe) is 1.2 years. In comparison to the highest irradiation in Africa, which is more than 2,700 *kWh/m²/year* (Table 8.1), the EPBT can be as short as 0.7 years. Assuming the solar panels have a 25-year lifetime, this means that a solar PV system in Africa can produce as much as 35 times the energy that was needed to produce it.

Example 8.1: *Power Density of a Large-Scale Solar PV*

A large solar PV array located in the Mojave Desert, California is documented in Rever (2017). The huge plant of 550 *MW* covers an area of 16 *km²*. Assuming that all the area is covered with solar panels it means that the power production from the solar modules is 34 *W/m²*.

8.3.3 Temperature dependence

As well as the irradiation the solar module output is dependent on the temperature. Under standard testing conditions the module temperature is assumed to be the same as the ambient temperature. However, under normal conditions the solar cells are heated and will give out a lower voltage output when they are subjected to heating. A typical PV module will convert only 14–20% of the sunlight into electric power; the rest of the sunlight is converted into heat. The resulting operating temperature will be at a point between the heat loss to the surroundings and the heat generated in the module.

Heating of the solar panel can result in a power loss of around 8% in a year. This of course means that less energy will be produced by the PV system. It can also reduce the PV system's lifespan. The temperature dependence is expressed as the relationship between the short circuit current and the open circuit voltage versus the temperature, respectively. The temperature coefficients indicate how much the voltage and current change for each degree increase from the standard 25°C reference.

The open circuit voltage will typically *decrease* by 0.4% for every degree K deviation from the reference, while the short circuit current will *increase* by 0.05% for every degree increase. Since the voltage depends more on temperature than the current, the power will decrease as the temperature increases.

> Solar power output decreases with increasing temperature.

Example 8.2: *Power Requirement for Irrigation Pumping*

Temperature will influence the required solar power for pumping. The required hydraulic energy is multiplied by a temperature factor

$$\frac{1}{1 - \alpha(T_{cell} - T_0)}$$

where α is the PV module temperature coefficient, which is of the order 0.45%/°C (Campana *et al.*, 2016). T_{cell} is the solar cell temperature and T_0 a reference temperature of 25°C. So, with a cell temperature of 50°C the required power will increase with around 13%.

Global warming because of climate change may increase the risk of lower efficiency, since the PV cells operate best at around 25°C.

Efficiency loss caused by increasing temperature has been studied in Saudi Arabia (Adinoyi & Said, 2013). A module temperature increase from 38°C to 48°C resulted in 10.3% losses in efficiency. This corresponds to a temperature coefficient $\alpha = 0.9\%/°C$ (now based on the reference temperature 38°C). The annual loss of energy due to high temperature was estimated at around 10%.

8.3.4 Floating PV systems

In some areas of the world, notably in China, Japan, South Korea, India, the UK, France, Italy, Brazil, Portugal and the U.S., there has been a move towards floating solar panel technology, in which the panels float on the surface of water. This will have some appealing consequences including that the panels do not take up valuable space on land. Japan has implemented more than 55 MW_p of floating solar across 45 plants. This includes the floating solar panels on the

Yamakura Dam, the world's largest floating PV project (see Chapter 13). The water cools the panels, which will make them perform more efficiently. Since elevated temperatures can reduce the lifespan of the panels, they have a longer lifespan than ground-mounted panels. Another important secondary effect is that the panels prevent evaporation from the water surface. Since evaporation is a major problem in many water reservoirs (Olsson, 2015, Chapter 10) in water-scarce areas, it has been suggested that covering the dam with solar panels partially solves this problem.

8.3.5 Technology development

As indicated, in coming years there will almost surely be an emphasis on developing solar cells with higher efficiency, improved thermal stability and a longer PV module lifetime. There is also an interesting move towards raising the energy efficiency by capturing the light from the rear side of the solar cell, and to capture diffused light. As indicated, the conversion efficiency will exceed 20% within a few years and the degradation will be as low as <0.3%/year (WEC, 2016, Chapter 8).

8.4 SYSTEMS OF SOLAR CELLS

A single PV cell is usually quite small and may typically produce 1–2 *W* of power. By connecting solar cells in series and parallel large *modules* are formed that can produce a higher power output. Typically, 36 cells are connected in series to form a module. There are many different types of PV modules and their structure depends on the type of application. Then the modules can be further connected to larger units called arrays. Therefore, typical solar PV systems are built up from a number of modules and arrays.

A single solar cell provides a voltage of around 0.6 *V*. By connecting several cells in series, a higher voltage will be produced. It is common that modules are designed to be able to charge a 12 *V* battery. Thirty-six cells in series provides an open circuit voltage of 21 *V*. Depending on temperature reductions and best operating conditions the module may produce 17–18 *V*. Since it will only require around 15 *V* to charge a 12 *V* battery, this will be sufficient. The voltage may well drop due to less light or further voltage reduction in parts of the system. More on batteries can be found in Chapter 10.3.

The current from a module is a function of the size of the solar cells and their efficiency. A single solar cell has an area of typically 100 cm^2. As a rule of thumb, the current from a module will be of the order 3.5–4 A. Having 36 cells in the module means that the area is 3,600 cm^2, or a square of $60 \cdot 60$ cm^2.

The most common PV system consists of flat-plate cells. Another more advanced category is systems where the sunlight is concentrated. Concentrating optics are used to focus the light on small solar cells. This increases the power output and decreases the number of required cells.

A complete solar PV system needs more than the solar cells to provide the output electric power. It includes a charge controller, an inverter (see 4.5) and batteries that provide adequate electric power. Pumps are usually powered by alternating current (AC) motors, which are typically more robust and cheaper than direct current (DC) motors. Since the solar panels produce DC currents a DC/AC converter is required for the AC pumps.

Energy storage is a concern (further examined in Chapter 10) and batteries are used to store the PV output in order to smooth or sustain the system operation when the solar irradiation is insufficient. Control system aspects are illustrated in Chapter 11.

8.5 ENERGY REQUIREMENTS FOR WATER OPERATIONS

Example 8.3: *Solar PV Output for Pumping, Eastern India*

The actual power production of a solar array over a day for different months in eastern India has been measured by Rahman and Bhatt (2014). Since the solar irradiance varies with the length of the day, the DC power output from a 3 kW_p solar array in the Patna region was measured every 15 minutes on bright sunshine days in different months of the year. The array tracked the sun with a manual mechanism. From 9:00 to 14:30 in almost all months, except November to January, the power rating of the array was within the interval range 1.9–2.4 kW. As a result, a 2.2 kW (3 hp) pump could be operated at close to rated power for six hours daily. So, a rule of thumb for eastern India is that a solar array of 1 kW_p can provide sufficient power for a 1 hp (0.74 kW) solar pump.

Example 8.4: *Small-Scale Solar PV for Desalination, Jordan*

This solar PV desalination plant is in Jordan and is designed to produce fresh water from brackish water, as reported by Hoffman (2017a). The plant delivers 95 kWh per day to produce 22 m^3 of fresh water from 37 m^3 of brackish water. This gives an energy requirement of 95/22 = 4.3 kWh/m^3 to produce fresh water (compare 5.3.5).

8.6 FURTHER READING

Varadi *et al.* (2018), Chapter 2, present a pedagogic introduction to solar power technologies. An easy-reading introduction to the various technologies is also found in WEC (2016, Chapter 8). Goswani (2015) contains a comprehensive description of major developments in solar energy. A case study on rural electrification in Papua New Guinea is presented by Kaur and Segal (2017).

White (2014) is an easy-read introduction into solar PV, while White (2016) is a comprehensive text on solar PV and installation.

ABB has published an ambitious technical application paper on solar PV systems (ABB, 2018) that explains their principles, connection, mounting and economy.

SAM (System Advisor Model) is a detailed performance and financial modelling tool for solar and wind energy applications (https://sam.nrel.gov/). SAM is based on computer models developed at NREL, Sandia National Laboratories, the University of Wisconsin and other organisations. SAM bases its calculations on an extensive and freely available database of meteorological and climate data for several locations worldwide. It provides the dynamic operation profile, on an hourly basis, of solar thermal generation, solar PV generation and wind energy and can therefore support the simulation of load performance of e.g. desalination systems. SAM contains online databases for both solar PV and wind: the NREL National Solar Radiation Database for solar resource data and ambient weather conditions as well as the NREL WIND Toolkit for wind resource data.

The site PV Education Network (http://pveducation.org), developed by Christiana Honsberg and Stuart Bowden at Solar Power Labs at Arizona State University (ASU) (http://pv.asu.edu/), contains a wealth of good illustrations and practical examples of PV generation and solar energy. Parida *et al.* (2011) present an overview of solar PV technologies.

The basic calculation for the dynamic energy yield of solar flat panels is presented in the form of Python code by Piani (2017).

Chapter 9

Wind

"First, there is the power of the Wind, constantly
exerted over the globe...
Here is an almost incalculable power at our disposal,
yet how trifling the use we make of it!
It only serves to turn a few mills,
blow a few vessels across the ocean,
and a few trivial ends besides.
What a poor compliment do we pay to our
indefatigable and energetic servant!"

Henry Thoreau, 1817–1862.

The global installed capacity of wind is expected to continue its rapid growth. Over the five years up to 2022, the cost of wind projects is predicted to drop by 14% as the industry continues along the learning or experience curve (Lienhard *et al.*, 2016). It should be noted that wind power technology needed quite a long time before it took off commercially to become a significant part of total power generation. More than 93% of the investments in onshore wind from 1983 to 2014 occurred after the year 2000. As shown in Figure 3.6, the installed wind power capacity is negligible in African countries compared to

© IWA Publishing 2018. Clean Water Using Solar and Wind: Outside the Power Grid
Author: Gustaf Olsson
doi: 10.2166/9781780409443_0111

China, India, North America and Europe. Unquestionably there is a huge wind power capacity waiting to be utilised.

The fundamental properties of a wind turbine power are explained in 9.1. Like solar PV, the wind turbine produces power intermittently. Therefore, the wind capacity factor is important to know, as explained in 9.2. Intermittent production and the need for energy storage is the topic of Chapter 10, while economic and financing aspects for renewable energy are described in Chapter 12.

9.1 BASIC PROPERTIES OF WIND TURBINE POWER

The kinetic energy from the wind creates the rotational energy of the wind turbine. Then a generator converts the rotational energy into electric energy.

Obviously, the generated power of wind turbines depends on the wind speed. Theoretically, it is proportional to the cube of the wind speed. This means than when the wind speed is doubled, the wind power increases by a factor of eight. If the wind speed is too low, then the turbine does not produce any power and it may also stop producing power if the wind speed is too high. It is also a well-known fact that the wind speed can change within minutes, so the generated power from a single turbine is apparently quite variable. If several turbines are combined in a wind farm, then the variability can be reduced. Today it is common that turbines generate electric power at varying speeds, so the rotor is continuously matched to the wind speed. This will of course utilise the wind power more than if the turbine generated electric power only at a given rated speed.

When the wind speed doubles the wind power increases eight times.

The kinetic energy stored in a flow per unit volume is

$$E = \frac{1}{2} \cdot \rho \cdot v^2$$

where ρ is the density of the air. For a stream flowing through a transversal area S (here it is the area that is swept by the wind turbine) the flow rate is $S \cdot v$. The generated mechanical power P is determined by both the wind speed and the area S according to the so-called Betz formula:

$$P = \frac{1}{2} \cdot \rho \cdot S \cdot v^3$$

where ρ is the density of the air. The available energy in the wind is obtained by integrating the power during a time interval T_p, typically one year:

$$E_{annual} = \frac{1}{2} \cdot \rho \cdot S \int_0^{T_p} v^3 dt$$

The Betz formula explains clearly that the wind turbine power increases significantly not only with the circular area S but also with the tower height. At higher levels the wind is less disturbed by surrounding objects like hills and buildings. Also, wind speeds in the lower atmosphere increase with elevation and this is important given the v^3 dependence of extracted energy.

A wind generator requires a minimum wind velocity of 3–5 *m/s* and will deliver the nominal capacity at a wind velocity of 12–14 *m/s*. Wind turbines have been developed to have a favourable operating condition at 10–12 *m/s*. These wind towers have a larger rotor diameter per *kW* of power. At high speeds small-scale generators, for safety reasons, are blocked by a braking system. High-power wind generators turn the rotor blades to stop the rotation and the mechanical brakes are used only for gross failures. There are generators that can adjust to the wind direction so that the power output is constant.

The wind generator speed is variable since the wind speed is not constant. The AC power frequency delivered, however, should be constant. Therefore, the rotors of the generator are connected to inverters to control the voltage and frequency of the AC power delivered.

Globally wind power has been developing towards larger wind turbines and higher power outputs. The largest commercial turbines today (Vestas) have a power rating exceeding 9 *MW*. For our purposes of decentralised power sources outside the grid we will consider much smaller systems.

Not surprisingly, small wind power units have a higher capital cost per *kW* compared to large systems. For the considered off-grid applications we will focus on small wind turbines (SWT). There is a wide range of small-scale turbines, from 'micro SWTs' rated at less than 1 kW to 'midi SWTs' reaching 100 kW. They are commonly used as stand-alone electricity systems in off-grid locations. Obviously, if the location is not windy there is a lower load factor and a higher capital cost per kW. Another challenge is wind turbulence caused by obstacles in the surroundings. A high tower will reduce this turbulence but will increase the cost. Simply put, a turbine with a larger tower and a relatively larger rotor (compared to its maximum power output) will produce more energy per installed unit of capacity. The wind velocity on hills and ridges can be higher. On the other hand, the air density is lower at high altitudes. In cold climates the air is denser in winter, resulting in more wind energy.

A problematic issue is that solar PV is an increasing competitor to wind power, as solar PV cost has dropped so dramatically during the last few years. For example, wind power water pumping can compete with solar power water pumping only at high wind speed and low solar irradiation value (Campana *et al.*, 2015). To find the most cost-effective solution for irrigation depends on the location. Another disadvantage of wind compared to solar PV is that the wind tower machinery needs frequent maintenance and repairs, which can be problematic considering the sub-Saharan conditions (Varadi *et al.*, 2018).

9.2 WIND POWER EFFICIENCY

Wind speeds can change significantly in just minutes. The output of wind turbines is therefore variable. As already noted, the output grows with rising wind speed and is constant above the rated wind speed.

Since the wind turbine cannot run at nominal speed continuously the real delivery of energy is expressed as the capacity factor. A wind turbine located offshore normally has a higher capacity factor than an onshore turbine. For example, onshore turbines in the UK have 26% while offshore wind has 35% (WEC, 2016, Chapter 10). In Denmark, a pioneering country in wind power, the power plants were running for 20–25% of the time in the 1980s and 1990s. In 2017 the offshore 160 MWe wind power park Horns Rev operated for some 3,500 hours or 40% of the time. The typical operating range for offshore is now 3,500–4,000 hours per year or 40–46%. Today large offshore wind power parks have a capacity of around 50% and will probably increase to 55–60% once 10 *MW* turbines have been installed (Svensson, 2018).

<table><tr><td>The capacity factor for onshore wind power is of the order 25% while offshore can reach 50%.</td></tr></table>

It is apparent that the variability of wind power generation can be minimised if many wind turbines are connected in a large grid. When generation is connected over a large geographical area the local variations in wind can compensate for each other. As a result, the intermittency of wind generation can be reduced significantly. This means that off-grid wind turbines are more sensitive to changes in wind and local stand-alone wind turbines will have a relatively higher variability than a wind farm.

Wind energy technology is a mature technology and the systems operate at very high rates of availability. Availability of the turbine is the proportion of time the turbine is ready for operation. Onshore turbines typically have availabilities of 98%, while offshore turbine availabilities are slightly lower (95–98%). Today's turbines already extract nearly 50% of the energy conveyed in the wind.

Solar PV and wind can sometimes compensate for each other and create a less variable power production if they are connected. The wind may blow on a cloudy day or during the night, thus compensating for lack of solar power.

9.3 FURTHER READING

WEC (2016), Chapter 10 provides an easy-reading introduction to wind power. World Bank (2017) presents detailed information about global access to wind power.

IRENA (2015c) reports on technology quality for small wind turbines. A key issue is to limit the cost in developing markets but still maintain the reliability and quality of the systems.

For the interested reader the textbook by Manwell *et al.* (2011) gives an excellent introduction to wind power systems.

Chapter 10

Handling variable production

> "Nature is an endless combination and repetition of very few laws.
> She hums the old well-known air through innumerable variations."

Ralph Waldo Emerson, author, 1803–1882.

It is important to keep in mind that the maximum capacity of solar PV or wind turbines (expressed in kW or MW) is seldom achieved. Solar PV can only produce when there is sunlight and wind towers can only deliver power when the wind is blowing. From the user's point of view, the delivered useful energy (in kWh or MWh) is the interesting information. This is why the concept of "real" power and energy delivery is essential. Solar PV and wind energy sources have an intermittent production. Solar PV production can change in a few seconds if clouds or precipitation limit the availability of solar radiation for conversion to electric power. Wind speeds can change significantly in only minutes.

Unlike conventional power generating systems like hydropower, renewable production cannot be adjusted to demand. The power delivery from nuclear- or coal-powered thermal plants is predictable and is determined by the system operator. In contrast, renewable energy sources have a production completely independent of the demand. To adapt production to consumption some kind of energy storage is needed in order to compensate for short-term or long-term differences between production and consumption.

Often, however, peak production times for solar and wind energy align with peak demand periods for water pumping and treatment. If there is an excess power production the extra energy can always be put to good use, for example to store as a fresh water buffer, as discussed in 10.1. However, electric energy also must be available when the solar PV or wind cannot deliver. Then electric energy storage capacity is needed. Various storage technologies are summarised in 10.2. The most commonly used is battery storage. Some battery technologies are described in 10.3 and characterising parameters are presented in 10.4. Hydrogen storage (10.5) is another possibility with a huge potential for long-term storage. A third possibility is to use excess energy to pump water to a high elevation, like pumped storage hydroelectricity, shown in 10.6. It is always possible to use traditional diesel generators as a last resort if both solar and wind fail to satisfy the load, as shown in 10.7. The cost of energy storage is of primary importance, discussed in 10.8.

One thing should be remembered: production being variable and intermittent does not mean that it is unpredictable. Depending on the economic resources available, the production of both solar PV and wind can often be predicted quite accurately at least for the next 24 hours. Furthermore, it should be kept in mind that most electric energy generation does have some downtime.

10.1 INTERMITTENT PRODUCTION CHARACTERISTICS

There are two key parameters that can characterise variable production from renewable energy sources: capacity factor and load profile. Both have consequences for water supply operations.

10.1.1 Capacity factor

The capacity factor is a measure to quantify the total energy generated from a specific source. The variability of wind and sunshine is the reason that the wind power or solar PV cannot be used at full capacity at all times. The net capacity factor of a power plant is defined as the ratio of its actual output over a period of time to its potential output if it were possible for it to operate at maximum power output continuously. Wind capacity factors range from 20% to around 50% so the power density will be reduced accordingly. For solar PV the capacity factor depends on the latitude and weather pattern. Some typical values are 9% (UK); 13–15% (Massachusetts, US); 18% (Portugal); 19% (Arizona, US).

It ought to be mentioned that the capacity factor is also below 100% for conventional power systems. For hydroelectricity, the global average is 44% (Kumar *et al.*, 2011, p. 446) and the range is 10–99% depending on design and local conditions. The averages of the continents vary from 32% (Australia, Oceania) to 54% (Latin America). In the US the downtime for coal-powered thermal plants is around 12% and for nuclear plants 10%.

Of the three approaches for reconciling electricity supply and demand, energy storage gets most of the attention and most of that interest is focused on batteries. The electric vehicle market has led to sharp gains in cost-performance of battery systems – a trend expected to continue. The use of hydrogen for energy storage is getting increasing attention, which should be of major benefit for RE systems. Batteries for energy storage are considered in 10.3–10.4 and hydrogen storage in 10.5.

There's a lot of confusion about how much storage capacity is actually needed, and what cost targets the storage must meet. Energy storage depends on the load profiles for each individual application and on consumer requirements.

10.1.2 Load profile

The most efficient way to use renewable energy, from solar or from wind, is to feed the power directly to the load. Due to the intermittent nature of the production, it does not always fit the load profile. Lighting is the best example of the mismatch between generation and load – the load is turned on when solar PV generation is not available. On the contrary, cooling is an apparent example of a load that follows the availability of the solar resource: the more sun, the hotter the weather and the higher the demand for cooling. Commercial and domestic loads usually follow quite strict timings, for example work hours and times of meal preparation.

It is obvious that some energy storage is required to compensate for the mismatch between production and consumption. To deliver the power via storage will always reduce the efficiency of the system. Naturally storage adds cost to the system as well.

> Energy storage is required to compensate for the mismatch between production and consumption.

The simplest load profile is a constant load over time. However, most load profiles follow the daily living pattern. To estimate the load profile several factors ought to be considered:

- The day of the week,
- The hour of the day,
- The outside temperature,
- The season,
- The service demands.

Considering the load profile on a household or village level in a rural area in Africa or developing Asia will emphasise special needs. Figure 10.1 illustrates a typical daily load profile for households in rural areas. Similar load profiles from Zimbabwe and Uganda are published by Prinsloo *et al.* (2016, Figure 3). Note that the high load peak in the morning is just around sunrise and the evening peak is after sunset.

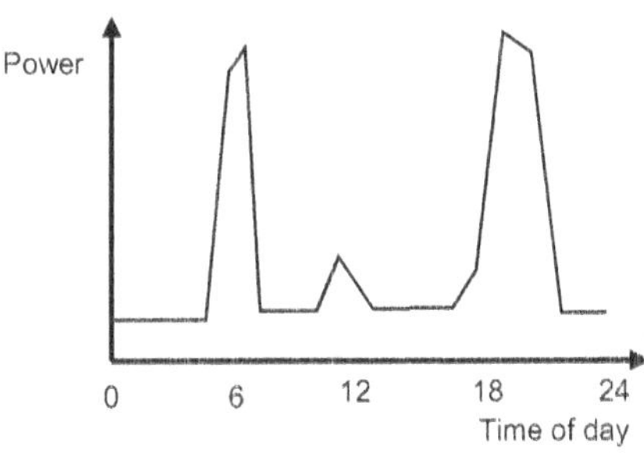

Figure 10.1 Typical household electric load profile in rural areas. Adopted from IEA (2013) and Prinsloo *et al.* (2016).

Institutions such as schools, health centres or community centres have a different load profile, since most of their activities take place during typical working hours, usually in the daytime. Since the highest power demand occurs during sunshine hours, a significant share of the PV-generated electricity can be used immediately when generated. The peak load and daily energy demand can vary significantly, depending on the specific user of the system. It may, for example, include running refrigerators for cooling medicines at a health centre or charging batteries and mobile phones for villagers at a community centre.

A number of load profiles from various customers of a mini-grid in Tanzania are presented by Hartvigson and Ahlgren (2018). The authors had first estimated load profiles based on interviews, but later measurements of the load profiles showed quite different behaviour. The lesson learnt is that simplified load profiles may lead to inappropriate sizing of the power production. Predicted and measured load profiles for a household and for three small workshops with machine operation loads during the day are presented. The authors also emphasise the need for more data on load profiles in rural areas.

> Knowing the load profile is essential to find out the need for energy storage.

The renewable electric energy will not primarily be used for heating since the regions considered are relatively warm countries. However, in rural areas in Asia there is a need for heating for substantial periods of the year

The load profiles of water pumping and water cleaning can adapt to available electric power. Pumping water can be used as a balancing activity. When there is insufficient power supply pumping can be switched off and when power capacity exceeds the need pumping can be used to store energy in elevated water tanks.

Water supply treatment via desalination and disinfection can also run intermittently. This of course requires water storage capacity for untreated as well as for treated water.

10.1.3 Intermittent desalination

Intermittent use of a desalination plant to meet baseload water demand requires oversizing the plant relative to what would be needed under steady operation. Another difficulty is that the long-term reliability and the operational life of membranes are affected by intermittent operation and variable pressure (Lienhard *et al.*, 2016).

Intermittent power supply for desalination causes damage to the equipment. The membranes in a reverse osmosis plant are the most sensitive parts of the system. They are designed to have a constant pressure so they will last a long time, but with variable pressure their usual operational life of eight years (with a well operated plant) may reduce to two years. Research is needed to study and further understand the phenomena taking place within RO membranes operating under variable pressure and intermittent conditions.

Another important problem observed in transient operation of desalination systems is the increased risk of membrane scaling, bio, organic and mineral fouling, collectively known as fouling. Increased fouling also necessitates greater cleaning activities and the membranes may need to be replaced earlier (Lienhard *et al.*, 2016, Chapter 2).

> Intermittent power supply for desalination is shortening the operating life of the equipment.

A desalination plant that requires a certain production around the clock will need power for 24 hours, but the solar PV is available for only around eight hours/day. Thus, the power of the solar PV system must be at least three times the size of a power plant that operates around the clock.

> To produce a certain amount of energy, a solar PV system operating eight hours per day will require a power rating three times higher than a system operating around the clock.

10.2 STORAGE OF ENERGY

A solar PV system is limited to working only in daylight unless it is integrated with an energy storage system. A solar thermal system with a medium for thermal storage can operate through the day. Heat can be stored during the day and excess daytime heat can potentially be stored, as shown in Chapter 6.

Often solar PV and wind can complement each other to provide a more reliable power source (Weiner *et al.*, 2001). Meteorological data will decide whether the accumulated wind and solar energy production can satisfy the power demand.

> In any intermittent production the need for storage capacity must be well thought out.

The combination of solar, wind and battery storage is often too costly for poor areas that may be satisfied, at least primarily, with less ambitious energy supply.

10.2.1 Storage requirements in low-income versus high-income countries

The priorities concerning variable energy sources appear to be quite different in high income and in low-income regions. In the former case power availability around the clock is emphasised, even if it means a higher cost for control and storage. For the user in a low-income country there is a higher acceptance that power is not available around the clock, particularly for those people who previously had no electric power.

10.2.2 Storage technologies

There is no "one-size-fits-all" solution to storage need. There are several different technologies available and the type of need as well as economy will determine the kind of technology to be used.

Different technologies are suitable depending on the timescale:

* *Timescale of minutes to less than an hour:* The purpose of the storage is to keep the power delivery from solar and wind smooth. Also, short variations in the consumption can be compensated for.

- *Diurnal variations*: The solar power has a diurnal variation, so the cycle time is one day. To fully compensate for variations in this timescale the buffer storage capacity needs to be several *kWh* per *kW* of renewable energy capacity.
- *Bridging periods of bad weather*: This is a period where the skies are overcast and there is hardly any wind. To store capacity during such a period is naturally much more challenging than with diurnal variations. To afford this storage capacity may be unrealistic in a low-income region.
- *Seasonal variations*: In a cold climate there is a large temperature difference between cold and warm seasons. Thus, the need for energy storage for the winter is high. On the other hand, the difference between the coldest and warmest month in a tropical or subtropical country is much less. Therefore, the need to store energy for the coldest month is not as great.

The most important storage technologies are listed here. They are suitable for different time and energy scales and not all of them are appropriate for a low budget.

- *Batteries*: For short variations it is common to use batteries. Car lead-acid batteries have been used for decades and are a viable alternative in low-income countries. However, the requirements for the battery for a renewable energy system are different than for a car. Battery storage is described in 10.3–10.4. The discharge time is several hours.
- *Flywheels*: The electric power can be stored in a flywheel storage. In other words, the electric energy is converted into kinetic energy, coming from a motion of a spinning mass, or a rotor. Most modern high-speed flywheel energy storage systems consist of a rotating cylinder, supported by magnetic bearings. The flywheel operates in a vacuum to reduce drag. A motor-generator unit will convert the electric energy to rotation; and to utilise the stored kinetic energy the motor-generator unit is used in the opposite direction to generate electric power. Flywheel storages are common in aerospace and telecommunications. Typical timescales for discharge are in the range of seconds to minutes.
- *Compressed air*: A compressed air system uses excess electric energy to compress ambient air and store it under pressure in a

pressure container or an underground cavern. When electricity is needed the pressurised air is heated and drives a turbine that is connected to a generator for power production. The discharge time can be several hours, but the efficiency is quite low.

- *Pumped storage*: Using excess power to store water in a tank at an elevated level. Pumped storage can be used at any scale, from household level to large-scale utility level. The discharge time is typically from several hours to a day.
- *Hydrogen*: Electric energy can be converted to hydrogen by electrolysis. The stored hydrogen can be converted back to electricity, but the overall efficiency is quite low, around 30–50%. Hydrogen storage may have an increasingly important role to play, since the energy density of hydrogen is higher than in most other storage technologies. However, hydrogen technology is probably not relevant for low-cost installations.
- *Diesel-powered generators*: To manage longer-term storage there is still a possibility in some areas of using traditional diesel-powered generators.

The storage technologies are at different development stages. Some of them are mature and proven over a long time, like pumped hydropower. Flywheels are getting more common in certain applications. Battery technology has been developed over many decades and lead-acid batteries are used everywhere. More advanced battery technology is being developed at a high rate, driven by the growth in development of electric vehicles.

For lighting and clean water operations we are considering the following alternative storage technologies:

- Batteries,
- Hydrogen,
- Pumped storage, and
- Traditional diesel generators.

10.3 BATTERY STORAGE

Globally, battery storage for large-scale as well as small-scale applications has grown rapidly over the last few years. Grid-connected stationary battery storage capacity has grown from 345 MW in 2010 to

620 *MW* in 2013 and 1,719 *MW* in 2016. In the year 2016 global battery capacity grew by 50% (REN21, 2017a).

This rapid battery development, together with low-cost solar PV systems, is seen by many experts as a potential disruptive technology that could dramatically change future energy markets.

The number of electric vehicles is increasing and as a result automotive companies are expanding into energy storage research and development. Tesla has released the Powerwall battery and companies like Mercedes-Benz and BMW are now marketing stand-alone batteries. This development will hopefully have a positive influence on the solar energy market.

Not all batteries are created equal, even batteries with the same chemistry. Batteries can be designed either to be high-power or high-energy and are often classified as one or other of these categories. We concentrate on three categories of batteries: lead-acid, lithium and saltwater. Lithium batteries have the longest lifetime, saltwater shorter and the lead-acid batteries the shortest lifetime.

Batteries are often evaluated according to their energy density (*kWh/litre of battery*) or total weight. This is important in transportation and in many consumer products like mobile telephones. By this measure lithium batteries have so far been the most successful. However, their cost is relatively high. A stationary battery can be assessed in a different way: even if its energy density is less the battery may still be successful; the total weight is not critical, so other properties can be considered. Consequently, a lot of attention is directed towards flow batteries and saltwater batteries.

Stationary batteries should be evaluated differently from batteries for transportation.

10.3.1 Lead-acid batteries

The cheapest and most common battery is a standard car battery of the lead-acid type. These batteries are a proven technology that has been used in off-grid energy systems for decades. This is probably the least expensive alternative available for households. A drawback of deep-cycle lead-acid batteries is their expected short lifetime. Even if they are carefully managed their lifetime may be only three years, which is much shorter than other components in the energy system. If the capacity of discharge is limited to 20% then the batteries may last five years.

There are three main types of lead-acid batteries: flooded, AGM (absorbent glass mat) and gel. In a flooded battery the electrolyte is in liquid form (sulphuric acid diluted in water) while in the AGM and gel the electrolyte cannot flow. The liquid in the flooded battery must be maintained, while the latter two do not need maintenance. They can also be used in any orientation.

Traditional car lead-acid batteries are usually not the best option for solar energy storage. Unfortunately, they are the type of batteries most commonly used in the developing world as they are cheap and easily available. A car battery is designed to produce a high current over a short time in order to provide power to start an engine. Typical loads powered by solar energy, on the contrary, usually require low currents over an extended period. The problem is not the battery chemistry; it is the design of the batteries. A traditional car battery will not survive for long if it is discharged more than 50%. Deep-cycle batteries are better for solar energy uses as they have larger electrodes than car batteries.

> Traditional car lead-acid batteries are usually not the best option for solar energy storage.

All three lead-acid battery types are available for deep cycle. A flooded battery has a price of around 70–80 USD/*kWh* while AGM and gel batteries are more expensive, around 260–330 USD/*kWh*.

At first glance the use of a car battery for the storage of solar-generated power may look more attractive than other options. However, in the long run this choice will almost certainly turn out to be wrong.

10.3.2 Lithium batteries

We use lithium batteries in our mobile telephones. Such batteries use some form of lithium chemical composition. They are lighter and more compact than lead-acid batteries. They can be discharged more, are robust and have a longer lifespan. They are more expensive; but the price of lithium batteries has dropped by nearly 20% per year since 2010. Lithium represents the key limiting resource for most batteries. The largest reserves of lithium by far are found in Chile, ahead of China, Argentina and Australia.

> Lithium batteries are lighter, more robust and more compact compared to lead-acid batteries. They can be discharged more and have a longer lifespan, but are more expensive.

The two main types of lithium batteries used in solar applications are lithium cobalt oxide and lithium iron phosphate. Today the cheapest lithium battery is the Tesla Powerwall, with a cost of around 500 USD/*kWh*. Kittner *et al.* (2017) explain that R&D investments in energy storage projects have been remarkably effective in bringing the cost per *kWh* of a lithium- battery down from 10,000 USD/*kWh* in the early 1990s to a trajectory that could reach 100 USD/*kWh* in 2019. This means that storage prices are falling faster than solar PV or wind technologies.

10.3.3 Saltwater batteries

There are interesting new types of batteries still in a development phase. Saltwater or sodium-ion batteries have attractive properties. They do not contain heavy metals but instead use saltwater electrolytes, consisting of sodium sulphate in an aqueous solution. This means that a saltwater battery can be easily recycled, in contrast to a lead-acid or a lithium battery. Today there are battery stacks available for 48 *V* or 24 *V* with a current rating of 82 *Ah* (see below) and above 2 *kWh* energy capacity. An interesting feature is that these battery stacks can be connected in parallel even if the discharges and voltages of the individual battery stacks are not the same. This cannot be done with lead-acid batteries.

Saltwater batteries are interesting not only from a performance point of view but also due to their price. This kind of battery is maintenance-free; even discharging deeply does not have a large impact on its performance. The price is of the order 400–500 USD/*kWh*. A saltwater battery can have up to 3,500 cycles with a 90% discharge. This corresponds to an operational life of almost ten years.

One drawback of the saltwater battery is that it is slightly heavier than a lead-acid battery. The other disadvantage is that it has to be charged and discharged at lower rates than the other battery types. A typical charge time is ten hours and discharge time 20 hours, because the current should not exceed around 14 A. The saltwater battery temperature should not exceed 40°C, which can be a problem in tropical regions. There are still very few manufacturers of saltwater batteries, so as competition increases and with economy of scale these batteries may become much more economically attractive.

10.3.4 Flow batteries

A flow battery, also called a redox flow battery is an electrochemical cell. Two chemical components dissolved in liquids are contained in the system and separated by a membrane, Figure 10.2. The chemicals provide chemical energy that is converted to electrical energy via the membrane. There is ion exchange through the membrane, while both liquids circulate in their own respective volumes. The ion exchange is followed by an electric current. The cell voltage typically reaches 1–2.2 V.

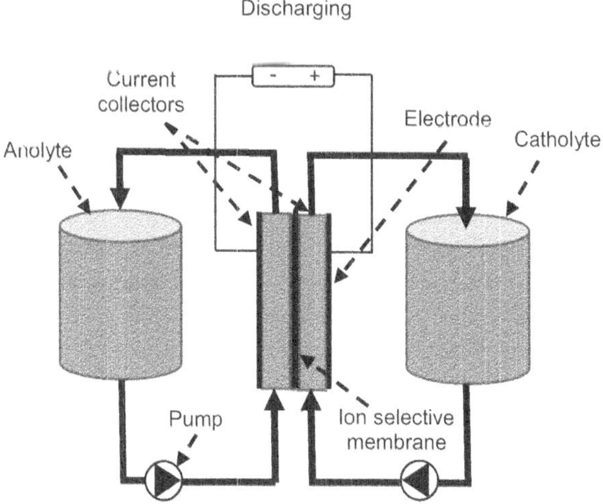

Figure 10.2 Principle of a redox flow battery. Modified from Akhil *et al.* (2013).

The battery can serve as a rechargeable battery, where an electric power source drives regeneration of the fuel. One of the biggest advantages of flow batteries is that they can be almost instantly recharged by replacing the electrolyte liquid. The energy capacity is a function of the amount of liquid electrolyte and the power a function of the surface area of the electrodes.

The fundamental difference between conventional batteries and flow batteries is that energy is stored as the electrode material in conventional batteries but as the electrolyte in flow batteries. A wide range of chemistries and electrolytes have been tried for flow batteries. Vanadium is increasingly being embraced by battery manufacturers as a core material in the production of batteries to be used in both small-scale and large-scale applications. The electrolyte is composed of vanadium salts in sulphuric acid. Vanadium redox-flow batteries (VFBs) have started to grow in influence as energy companies look to improve energy storage.

During a VFB battery charge, V^{3+} ions are converted to V^{2+} ions at the negative electrode through the acceptance of electrons. Meanwhile, at the positive electrode, V^{4+} ions are converted to V^{5+} ions through the release of electrons. Both of these reactions absorb the electrical energy put into the system and store it chemically. During discharge, the reactions run in the opposite direction, resulting in the release of the chemical energy as electrical energy.

Due to their superior performance and gradual price reduction lithium batteries have until now been the preferred technology. However, the demand for flow batteries has increased rapidly over the last couple of years. The flow batteries have low variable costs (USD/kWh) and use a wide state of charge (SoC) range. On the other hand, efficiency is lower than for the lithium batteries and fixed costs (USD/kW) are higher.

Activities in flow batteries in China illustrates the growing interest in using flow batteries for energy storage. Two recently announced projects in Hubei and Dalian are sized at 100 MW/500 MWh and 200 MW/800 MWh respectively. The Chinese government is aiming for a long-term strategy to push energy storage to integrate renewable energy.

10.4 BATTERY PARAMETERS

A battery can be characterised not only by its energy capacity but also by its discharge rate. Both these parameters are crucial for a storage battery used in renewable energy systems.

10.4.1 Battery capacity

A battery's capacity is the total amount of energy (*kWh*) that it can store. Capacity is often expressed in ampere-hours (*Ah*). The voltage is assumed to be given, so the capacity in *kWh* can be calculated. This is the discharge current a battery can deliver over time. By stacking several batteries, the total capacity can be increased. It should be noted that capacity does not indicate how much power a battery can deliver at a given moment; therefore, its power rating (measured in *kW*) should be declared. Consequently, a battery with a high capacity and a low power rating would deliver a low amount of electricity for a long time. A battery with a low capacity and a high power rating could run a high load but only for a relatively short time.

The electrochemical battery has the advantage over other energy storage devices in that the delivered power stays high during most of the discharge and then drops rapidly as the charge depletes.

Usually it is not possible (or advisable) to use all the capacity of a battery. Most batteries need to retain some charge at all times due to their chemical composition. The depth of discharge (DoD) of a battery refers to the amount of a battery's capacity that has been used. Most batteries have a recommended maximum DoD for best performance. For example, if a 10-*kWh* battery has a DoD of 90% then only 9 *kWh* can be used before recharging it.

The rating of a battery is usually expressed in amp-hours (*Ah*). The standard rating is a discharge current taken for 20 hours. Thus, a 100 *Ah* battery can discharge 5 *A* for 20 hours, from 100% state-of-charge to the cut-off voltage. It is important to know that this is not a linear relationship. If instead the discharge current is 100 *A* then the discharge time is less than one hour.

Battery capacity can also be expressed in terms of "energy capacity". This is the total energy (in *kWh* or *Wh*), expressed as the discharge power times the discharge time. The time is defined as the time for 100% state-of-charge to the cut-off voltage, which is the voltage that generally defines the "empty" state of the battery. Like capacity, energy decreases with increasing discharge current.

Temperature will significantly influence the battery. Since a lead-acid battery is an electrochemical device, an increasing temperature will accelerate the chemical activity while a colder temperature will slow it down. For a lead-acid battery an increasing temperature has several consequences:

- Increases performance,
- Increases internal losses,
- Decreases cell voltage for a given charge current,
- Shortens the battery life,
- Increases maintenance requirement.

10.4.2 Battery sizing

Assume that the power requirement is 300 W for 16 hours. This corresponds to an energy requirement of 4,800 Wh or 4.8 kWh. Assuming a battery with 12 V, this means that the required rating of Ah is 4,800/12 = 400 Ah. However, the battery's losses should be taken into consideration. Assuming 15% losses (85% battery efficiency), this means that 400 Ah will be divided by 0.85 to find the necessary rating. Then the battery cannot be discharged 100%. Assuming an allowable discharge of 60% means that the capacity will be further divided by 0.6 to obtain the required capacity. The final battery rating then has to be 400/(0.85 · 0.6) = 784 Ah or around 800 Ah. Typical rating of a lead-acid battery is of the order 100–600 Ah.

In summary, the required battery rating is calculated as

$$\text{Battery rating}\,(Ah) = \frac{\text{Total battery energy capacity}\,(Wh)}{\text{Battery efficiency} \cdot \text{Depth of discharge} \cdot \text{Battery voltage}\,(V)}$$

10.4.3 Battery classification

Batteries are classified in C- or E-rates, which is a way of showing the discharge current. It its normalised against battery capacity. A C-rate is a measure of the rate at which a battery is discharged compared to its maximum capacity. The rate 1C means that the discharge current will discharge the battery in one hour. For a battery with a capacity of 100 Ah it means that the discharge current will be 100 A. A C/2 rate would be 50 A. Applying a smaller charge/discharge rate will prolong the life of a chemical battery, so a discharge rate of for instance <0.1C is quite favourable for the battery's life. Similarly, using a battery discharge rate >1C is called a "high current" use. Usually a manufacturer specifies the maximum rating.

An E-rate defines the discharge power, so a 1E rate is the discharge power to discharge the whole battery in one hour.

Generally, the useful lifespan of a battery is 5–15 years, which is much shorter than the typical guarantee time of 25 years for solar cells. However, there is a lot of effort currently being devoted to battery development. Two particular demands are causing this: the rapid expansion of renewable energy and the increasing interest in electric vehicles.

10.4.4 Battery charge controller

A solar charge controller must match the voltage of the solar PV array and the batteries. It is usually defined by the current (ampere) and voltage (V) capacities. The charge controller need to have sufficient capacity to handle the current from the PV array. A standard practice is to design the solar PV charge controller 30% larger than the short circuit current (I_{SC}) of the PV array (see Chapter 8.2).

The charging voltage must be higher than the battery voltage for current to flow into the battery. There are two basic ways to charge a lead-acid battery:

- *A constant voltage* is applied across the battery terminals. As the voltage of the battery increases the charging current tapers off. This method requires only simple equipment, but is usually not recommended.
- *Constant current charge*: an adjustable voltage source maintains a constant current flow into the battery. This requires a more sophisticated charge controller but is the more commonly recommended technique.

10.5 HYDROGEN ENERGY STORAGE

Hydrogen as an energy storage system is receiving increasing attention and some systems are already in use. The systems are probably still too expensive for low-income countries but there is a huge research effort into hydrogen energy storage technology, so it should be mentioned here as a realistic possibility for energy storage in renewable energy systems.

To use hydrogen for energy storage is not new. Dr Allan Hoffman has been one of the key scientists working in Washington DC to influence the development of renewable energy and the awareness of the water-energy nexus. In 1995 he served as Associate Deputy Assistant Secretary and then Acting Deputy Assistant Secretary for DOE's (Department of Energy) Office of Utility Technologies (OUT) in the Bill Clinton administration. This Office had responsibility for developing the full range of renewable electric technologies as well as hydrogen and energy storage technologies. In 1995 he wrote an OUT report: "I also believe that over this time period, hydrogen will emerge as an important energy carrier to complement electricity, given its ability to be used in all end use sectors and its benign environmental characteristics."

The main attraction of hydrogen is that excess energy can be stored for a long time, not only from day to night, but also between seasons. For example, at northern latitudes excess renewable energy can be produced during the summer. Battery power cannot supply energy for any longer than a few days. If the excess energy is stored as hydrogen, it can be reused as electricity during the cold winter. Prototype systems using this facility are already in use. Hydrogen is one of the most promising ways of dealing with longer-term energy storage.

> The main attraction of hydrogen is that excess energy can be stored for a long time and then reused as electricity.

Simply explained, excess electricity is used to electrolyse water into hydrogen and oxygen. The hydrogen is stored and later used in fuel cells that generate electricity.

10.5.1 Electrolysis of water

In electrolysis, water is split into hydrogen and oxygen using electricity. Using excess electric power from solar PV or wind does not create any greenhouse gas to produce the hydrogen. At the same time the cost for extra energy is practically zero. The electrolysis takes place in a device called an electrolyser. There are a lot of different configurations, but the principle and the key reactions are shown in Figure 10.3. Different electrolysers function in slightly different ways, mainly due to the different type of electrolyte material involved.

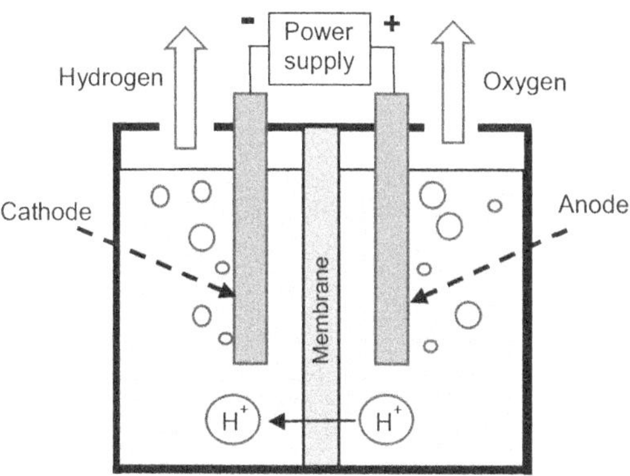

Figure 10.3 Principal electrolysis of water. Modified from Larminie and Dicks (2003).

When an electric current is passing through the water, electrons from the electric current cause an oxidation-reduction reaction. The electrons flow through an external circuit and the hydrogen ions selectively move across the membrane to the cathode. At one electrode, the cathode, electrons pass into the solution and cause a reduction. At the other electrode, the anode, electrons leave the solution completing the circuit, and cause an oxidation.

To carry out electrolysis the solution must conduct electric current. Pure water is a very poor conductor, so an electrolyte is added to enhance the water's conductivity. In most places, however, there are enough minerals in the water that the ionic strength or conductivity of the water is great enough for electrolysis. One problem with adding electrolytes is that they can electrolyse more easily than water.

At the cathode (the negative electrode), water dissociates into H^+ ions and OH^- ions. The H^+ ions are attracted to the cathode and are converted (reduced) to a hydrogen atom:

$$H^+ + e^- \rightarrow H$$

This is a highly unstable configuration, and therefore immediately reacts with another hydrogen atom to produce H_2, molecular hydrogen gas.

At the other electrode (the anode), oxidation occurs. The OH^- ions are attracted to the positive electrode, where they are oxidised to form oxygen gas (O_2) and hydrogen ions (H^+). The anode reaction is

$$2OH^- \rightarrow 4e^- + O_2 + 2H^+$$

Combining the cathode and anode reactions, we get the overall reaction:

$$2H_2O \rightarrow 2H_2 \text{ (gas)} + O_2 \text{ (gas)}$$

Producing hydrogen from solar PV is probably the cheapest method available today. However, it must be recognised that electrolysis, plus the associated equipment to compress and store hydrogen, is capital-intensive. The cost of equipment will probably come down, but the cost is currently still an obstacle.

Hydrogen does not need to be stored as a gas in pressure tanks. A new technology makes it possible to store the hydrogen in liquid form, a technology called Liquid Organic Hydrogen Carrier (LOHC). The technology is based on two separate processes: the loading (hydrogenation) and the unloading (dehydrogenation) of a liquid energy carrier. This liquid is composed of an organic molecule having similar physico-chemical properties to diesel. The LOHC-module consists of a hydration and dehydration unit and two tanks. One of the tanks contains the LOHC liquid carrier in its initial state. The hydration unit inserts hydrogen into the fluid in a chemical reaction. The loaded LOHC is then stored inside the second tank. Hydrogen can be discharged from the dehydration unit by an endothermic reaction (a chemical reaction that absorbs energy from its surroundings). The fluid returns to the initial tank.

A big advantage of hydrogen being chemically bonded to the liquid carrier is that it can be stored under ambient conditions (normal pressure and temperature) without suffering any self-discharge or the loss of hydrogen. One litre of the energy carrier can store an equivalent of 2 *kWh* thermal energy or, after reconversion, 1 *kWh* electrical energy.

10.5.2 Fuel cells

Hydrogen that is stored will sooner or later be converted to electricity. This is done in a fuel cell, a device that converts the chemical energy from a fuel into electricity. Here the fuel is hydrogen that will react with oxygen or another oxidising agent. In a battery the chemical energy comes from chemicals present in the battery. In a fuel cell there must be a continuous supply of fuel (hydrogen) and oxygen (usually from the air) to sustain the chemical reaction. Fuel cells can produce electricity continuously for as long as fuel and oxygen are supplied. They are not new inventions; the first was invented in 1838.

All fuel cells contain an anode, a cathode and an electrolyte: see Figure 10.4. There are many types, classified by the type of electrolyte and different start-up times.

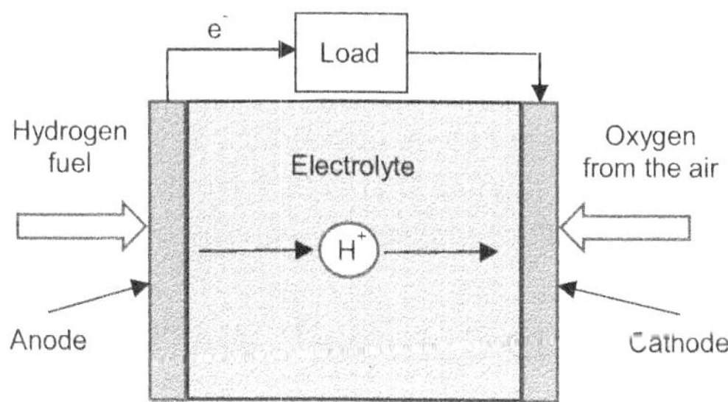

Figure 10.4 Electrode reactions and electric circuit flow for an electrolyte fuel cell. Note that the negative electrons move from anode to cathode, while the "conventional current" flows from cathode to anode. Modified from Larminie and Dicks (2003).

At the anode a catalyst breaks down the fuel (hydrogen gas) into electrons and ions. The catalyst speeds the reactions at the electrodes. The anode catalyst often consists of fine platinum powder:

$$2H_2 \rightarrow 4H^+ + 4e^-$$

The hydrogen ions travel through the electrolyte from the anode to the cathode. The electrons from the anode are led to the cathode via an external electrical circuit, thus producing a DC current. Oxygen from the air enters the cell at the cathode.

At the cathode another catalyst (often nickel) causes hydrogen ions, electrons and oxygen to react:

$$4H^+ + 4e^- + O_2 \rightarrow 2H_2O$$

where the water is considered to be a by-product. In other words, the summary reaction is very simple:

$$2H_2 + O_2 \rightarrow 2H_2O$$

A single cell will typically produce quite small voltages: 0.6–0.7 V. Therefore, several cells must be connected in series to create a sufficient voltage to meet the requirement for the specific load. If cells are connected in parallel a higher current can be supplied. Connection of cells is called a fuel cell stack.

The fuel cell will also produce water and heat as well as small amounts of emission, for example nitrogen dioxide (NO_2). The energy efficiency is not high; between 40% and 60%. This means that storage of energy as hydrogen causes significant energy losses, even though in many cases the waste heat can be utilised.

It should be emphasised that fuel cells have no moving parts. This makes them reliable sources of electrical energy.

10.6 PUMPED AND CLEANED WATER AS STORAGE

Storing energy as water in elevated places has been known about and practised for centuries. In hydropower water can be stored in reservoirs as potential energy that can be converted to electricity.

In many places pumped water storage is used to make use of excess energy. Then the potential energy of the water can be used to generate electricity when needed. This principle can also be used in stand-alone systems at the household level. Excess energy from solar PV or wind can be used to pump water to an elevated storage. Then the water can be used in two ways:

- If the elevation is sufficiently high, then the water may drive a turbine and generator to generate electricity;
- Otherwise the water can be used directly either for drinking (if the quality is adequate) or for irrigation, using only gravity.

The stored water can also be led into a desalination unit to be upgraded. Excess electric energy can also be used for pumping, desalination and disinfection.

10.7 DIESEL GENERATORS AS BACKUP

In many cases an existing diesel generator will be replaced by renewable energy. The old generator can be used as a backup for the intermittent generation, even if this will cost fuel and add to the carbon footprint. However, the advantage of using this backup is that even long-term lack of power can be rectified.

10.8 COST OF ENERGY STORAGE

To compare various technologies of energy storage, the concept of levelised cost of storage (LCOS) has been defined. This is usually expressed as a cost of energy storage capacity per kWh. Thus, the LCOS expresses the average cost over the life of the storage, including both capital and operational costs.

The huge solar PV and wind energy cost reduction has been achieved through an increasing mass production of PV panels and wind turbines. A similar revolution is apparently coming to energy storage. Worldwide there are massive investments in battery facilities. Energy storage cost has decreased from more than 500 USD/kWh in 2013 to around 200 USD/kWh in 2017 (Lazard, 2016). According to Boston Consulting Group (www.bcg.com) the cost is expected to drop below 100 USD/kWh in the next five to ten years.

10.9 FURTHER READING

There is a huge ongoing effort to develop batteries with larger capacity, smaller size and lower prices. To get acquainted with current storage research we refer to the Joint Center for Energy Storage Research (JCESR) in Argonne, supported by the US Department of Energy, and their webpage http://www.jcesr.org.

Sandia Laboratories has issued a handbook on electricity storage with a lot of technical information (Akhil *et al.*, 2013). IEC has published a white paper on electric storage systems (IEC, 2011); a major contribution comes from the Fraunhofer Institut für Solare Energiesysteme.

Delucchi and Jacobson (2011) discuss methods of addressing the variability of wind, water and solar energy to ensure that power supply reliably matches demand.

A comparison between lithium-based and flow batteries is presented by Uhrig *et al.* (2016).

Fundamental principles of fuel cells are explained in detail in Larminie and Dicks (2003). A very personal account of the development of renewable energy in the US is told by Hoffman (2016). He also regularly writes on energy issues on his blog (Lapsed Physicist, 2018).

For the reader particularly interested in renewable energy the book Jones (2017) is a deep well of information. The book concentrates on renewable power production and its integration into grids. Challenges related to variable production and energy storage as well as island power systems are analysed.

There is a lot of practical information on YouTube concerning battery choice, design, charging and discharging. Look for "solar batteries".

Chapter 11

Energy management systems

> "Automation does not need to be our enemy.
> I think machines can make life
> easier for men, if men do not let the
> machines dominate them."

John F. Kennedy, US President 1961–1963.

Any off-grid system for pumping and cleaning water needs a control system to take care of the energy management A solar PV unit or a wind turbine will produce electric energy independent of the required demand. An automatic control system will balance the power flow through the system. The energy production system includes not only the solar panel and the wind turbine; other components are inverters, batteries and charge controllers. The load consists of lighting, small household devices and low-power television sets as well as water pumps, disinfection lamps and desalination units. Their management must prioritise the demand side when the delivered power is not sufficiently high to provide any of the loads.

© IWA Publishing 2018. Clean Water Using Solar and Wind: Outside the Power Grid
Author: Gustaf Olsson
doi: 10.2166/9781780409443_0141

11.1 THE ROLE OF THE ENERGY MANAGEMENT SYSTEM

The role of the energy management system (EMS) is discussed. Since there is more than one source of electric energy, the energy production must be controlled. Likewise, not all the energy requirements of the loads can always be satisfied. This requires some prioritising between the various loads that are to be controlled by the EMS. The EMS can be implemented in a simple PLC.

The solar PV and the wind turbine are the key producers of energy. If they deliver less than minimum power, then the battery storage will be discharged. If this energy is insufficient, then as a last resort a diesel generator can be started.

The inverters (see also Chapter 4.5) will convert the DC power from the solar panels into AC power, while the wind turbine inverter will convert the generator power into a frequency that is suitable for the loads. It should be noted that the inverter rating must be large enough to handle the total power of the load that will be running at any one time. It is usually recommended that the inverter size should be 25–30% bigger than the total power of the load. For motor and compressor loads the inverter size should be at least three times the capacity of those appliances to be able to handle current surge during start-ups.

The battery charger (Chapter 10) will ensure that the voltage and current to or from the battery storage are within permitted limits.

The various loads are defined in the next section. The management system will also direct the power flow both when the solar and wind production is higher than the load and when it is insufficient.

A detailed design has to account for the changes in the efficiencies of the components depending on the load and the solar radiation and wind availability. The efficiencies are different if the system is operating in a PV/wind-to-load mode, PV/wind-to-storage mode or storage-to-load mode.

11.2 THE LOADS

The loads will be organised according to their importance. If the produced electric power from the solar panels, the wind turbine and the batteries is not sufficient to satisfy all the load requirements then some priority of the loads should be defined. The load with the highest priority (for example lighting) will be the last load to be shut down if energy production fails. The priority between the various loads will of course depend on the individual user. Naturally the water pump control is connected to the operation of the disinfection, desalination and the biological reactors.

The loads can be classified something like (see Figure 11.1):

- Lighting,
- Small appliances like radios, TV sets,
- Cooking facilities,
- Water pumps
- Disinfection lamps,
- Desalination unit,
- Biological water reclamation.

The energy management system will deliver available power to the loads according to the defined priority. Lighting, small appliances and cooking are probably the loads with highest priority.

Let us consider some operational modes and how to prioritise the loads:

(1) PV/wind power is greater than the required load.
   - Any excess power should be used wisely. If the batteries are not fully charged this ought to be the highest-priority load.
   - If the batteries are fully charged, then the excess power can be used to pump or treat water. Water serves as an energy storage, as described in 10.6.
(2) PV/wind power is not sufficient for the load requirements.
   - If there is sufficient storage capacity, then we are back to case 1.
   - If added storage power does not satisfy the load requirement, then we must define priorities for the loads. Only the most important loads should be connected.
   - If the batteries have been discharged, then the storage turns into a load, most likely the load with highest priority. Other loads must be disconnected according to their priorities.
(3) PV/wind power is not sufficient for the load requirements and batteries are discharged. This ought to be an unusual condition. Here the traditional diesel generator can be used as a backup, as pointed out in 10.7.

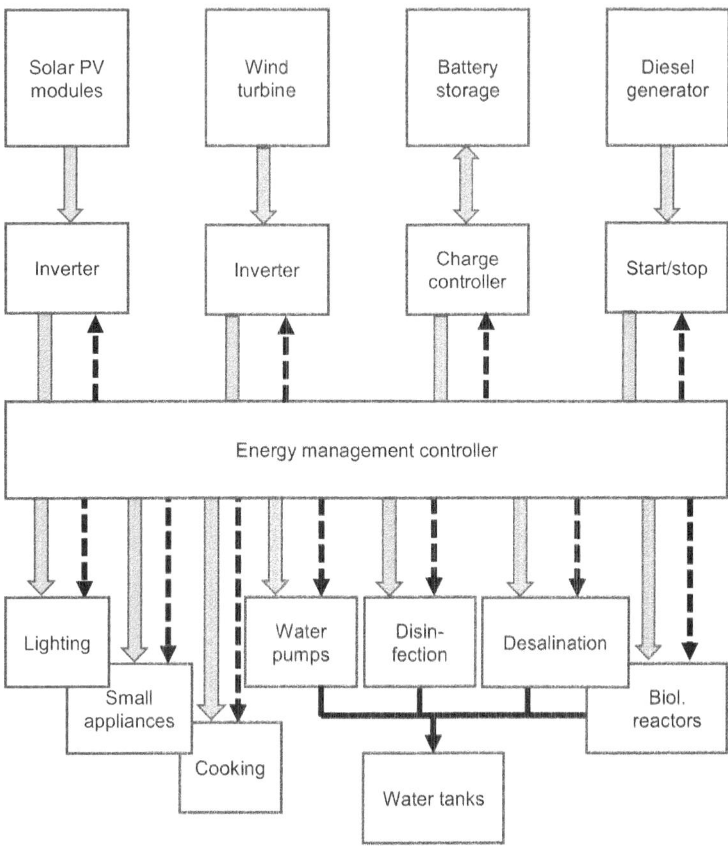

Figure 11.1 Block diagram of an energy management system for a combined solar PV and wind turbine system in a stand-alone operation. The production units are shown above the control system and the consuming devices are below the controller. The wide arrows indicate the energy flow through the system, while the broken arrows show the information and command flows.

It is apparent that energy management ought to take many operational modes into consideration. Naturally the user should not worry about the details. Still it is crucial to clearly explain to the user how the system prioritises its operations.

There are some issues to keep in mind with all installations of renewable energy:

- It is important not to confuse daylight hours with sun hours;
- Do not underestimate the power needed for the various loads. For example, a machine tool may require high power over short time intervals;
- When considering the load priorities, start with the small consumptions, but consider carefully the consequences of not being able to deliver power to a specific load;
- Water tanks can easily be supplied with a level detector so that the pump will be switched off when the tank is full;
- Solar panels are designed to run at certain temperatures. Remember that the capacity of the solar panel depends on temperature;
- Of course, a solar panel installation must be done properly, otherwise the calculated energy output may be quite wrong.

Some basic instrumentations for the water production system are:

- Pressure sensors;
- Level indicators for storage tanks;
- Temperature.

In more advanced installations the feed pump and high-pressure pump may be supplied with variable speed control, but fixed-speed pumps are more common.

Part IV

Applying Renewable Energy to Water Operations

To obtain clean water using solar PV and wind requires more than technology. Economising the systems is a key issue, discussed in Chapter 12. Even though costs for both solar PV and wind have decreased rapidly, there is a new kind of economy compared to traditional systems. The up-front capital cost is the dominating cost for renewables, while the "fuel" is free and maintenance much lower than for conventional energy systems. Thus, it becomes vital to find a financing system that takes the different cost structure into consideration.

In parallel with new technology there is a crucial need for education at many levels. The end user who has not previously enjoyed the opportunities that electricity can offer must be advised. Training at many levels of technical jobs is important if water operations using renewable energy are to be successful.

Energy systems require land areas. This issue is discussed in Chapter 13, and solar PV and wind energy land requirements are compared to land areas used for conventional energy systems.

Over the last few years a lot of experiences have been collected from using solar PV and wind for water operations. In Chapter 14 some cases illustrate how renewable energy has been applied to pumping and to desalination.

Chapter 12

Economy

"The dignity of each human person and the pursuit of the
common good are concerns which ought to
shape all economic policies."

Pope Francis' apostolic exhortation,
Evangelii Gaudium **(The Joy of the Gospel).**

Three aspects of economy are considered in this chapter. Costs for renewable energy are compared to fossil fuels in 12.1. Renewable energy is creating a huge job market. This also requires a corresponding effort in education and training, discussed in 12.2. In 12.3 some financing and payment options are reviewed.

12.1 COST OF RENEWABLES

Comparing cost of solar PV and wind with conventional power generation using fossil fuels requires that both capital cost and operating costs are considered. The costs to human health and the environment are often not taken into consideration when various energy production systems are compared, as commented on in Chapters 1–3.

12.1.1 Up-front capital cost versus fuel costs

In a renewable system the capital cost is the dominant cost, while the "fuel" for solar and wind is free.

A complete solar PV system consists of several elements to be connected with the solar modules. The solar cells usually represent the highest capital cost. The total cost for the system includes inverters (see Chapter 4.5), mounting structures and electric circuits. An important aspect is the cost of energy storage, in many cases a battery system, as demonstrated in Chapter 10.

Probably, all these components will be supplied by various vendors. It is crucial that the components are balanced for the best and most efficient operation. On top of this there are soft costs like a control system for power management, and labour costs for the installation.

With the dramatic decrease in the cost of solar cells it is obvious that the solar PV element of the system costs will be less dominant, so it is crucial to consider the total system cost.

The up-front capital cost is the dominant cost in a renewable energy system.

The capital cost is also the dominant cost for wind energy systems. For onshore wind the turbines make up the biggest portion of the capital cost. Turbine cost (including electric infrastructure and transportation) can represent a range of 64% to 84% of capital costs (IRENA, 2018). The main cost of operation and maintenance is related to the wind turbine itself, and is of the order of 50%. Unlike a solar PV the wind turbine has moving parts, which is always a more complicated operation from a maintenance point of view. Usually, turbine maintenance is done by the manufacturer, which also complicates the operation in remote areas.

12.1.2 Levelised cost of electricity

Levelised cost of electricity (LCOE) is way to express the lifetime costs divided by energy production and can be expressed in cost/kWh (see 1.4.5). The cost is defined as the present value of the total cost of building and operating a system over an assured lifetime. By using the LCOE concept different technologies can be compared, such as solar PV, wind, fossil fuels, nuclear etc., where the systems have different lifespans and project size as well as different capital costs, risks and capacities. Since the up-front capital cost does not take the whole picture into consideration the LCOE can help to make an informed decision to evaluate a project.

A simple LCOE calculation can be made using the expression

$$\text{LCOE} = \frac{\sum_{t=1}^{n} \frac{c(t) + om(t) + f(t)}{(1 + r)^t}}{\sum_{t=1}^{n} \frac{e(t)}{(1 + r)^t}}$$

where

$c(t)$ = capital expenditures in year t,
$om(t)$ = operation and maintenance costs in year t,
$f(t)$ = fuel costs in year t,
$e(t)$ = electric energy generation in year t,
r = discount rate,
n = lifetime (years) of the system.

The numerator in the equation is the total cost over the lifetime, while the denominator is the energy generated over the same time.

A significant attractive feature of renewable energy is of course that the fuel costs are free. The initial capital cost and the capacity factor (Chapter 10) are two critical parameters. The discount rate and the annual operating expenses are essential parts of the financing costs. There are LCOE calculators available online and for free, for example from NREL (National Renewable Energy Laboratory) in the US (https://www.nrel.gov/analysis/tech-lcoe.html). Note that there are important factors that are not included in the LCOE calculation, such as the projected utilisation rate.

12.1.3 Levelised cost for solar PV

The cost of the PV modules is determined by the raw material cost (type of semiconductor, metal frame, junction boxes etc.). The average LCOE of residential PV systems without any battery storage has been estimated at (WEC, 2016, Chapter 8):

0.38–0.67 USD/*kWh* in 2008
0.14–0.47 USD/*kWh* in 2014

The price reduction for residential PV installations (typically 0–4 *kW*) has been of the same order of magnitude in most western countries, as illustrated in Table 12.1.

Table 12.1 LCOE reduction for residential PV installations.

Region	Price Reduction %	Time Period
California	42	2008–14
Other parts of U.S.	52	2008–14
Japan	42	2008–14
Italy	59	2008–13
Australia	52	2010–14

Source: Data from WEC (2016, Chapter 8).

In regions where there is a lot of sunshine the LCOE is apparently favourable. Moreover, if the demand profile during the day and the sunshine hours are similar the need for storage capacity is smaller.

12.1.4 Levelised cost for wind energy

WEC (2016) has estimated that global onshore wind LCOE has decreased from 380 USD/*MWh* in 1983 to 70 USD/*MWh* in 2015 (in 2015 USD values). Continued performance improvements mean that LCOE for onshore wind was, in late 2017, at a record low 45 USD/*MWh* (Lazard, 2017). There is a wide span of LCOE for onshore wind, as shown in Table 12.2. The weighted averages are the largest in Africa, Oceania and the Middle East and the lowest in China.

Table 12.2 LCOE for onshore wind energy in some major regions.

Region	USD/*MWh*	USD/*kWh*
North America and Brazil	31–130	0.03–0.13
Africa, Oceania and Middle East	95–99	0.095–0.1
China	50–72	0.05–0.072

Measured in LCOE, onshore wind energy boasts some of the lowest electricity costs among renewable energy sources, as shown in Figure

1.2 (WEC, 2016). Because of the intermittent production in off-grid installations the cost for storage must be considered.

For our applications onshore wind power is the most interesting. This technology is reaching cost competitiveness against conventional power generation technologies (WEC, 2016, Chapter 10). Therefore, within the wind power profession most attention today is paid to technology developments in the offshore applications.

It is estimated that the global LCOE of onshore wind is likely to decrease up to 26% between 2016 and 2025, mainly due to lower installation costs, higher capacity factors and declining operation and maintenance costs. Higher hub heights, rotor diameters and turbine rating will account for most of the decline in the LCOE of onshore wind. It should be noted, however, that large-scale installations have the highest potential for cost reductions.

Wind projects are developed with an assumed operational lifetime of 20–25 years.

12.2 JOB OPPORTUNITIES

In interviews with experts on renewable energy they have been asked how many jobs will be needed for the renewable energy industry by 2050, given that 8.1 million are employed today (see Figure 12.1). Among the experts, 73% believe that more than 25 million jobs will be needed and 41% estimate that more than 45 million jobs will be required (REN21, 2017a).

> Millions of jobs will be needed in the renewable energy sector.

Utilities in sub-Saharan countries and in south Asia will in the future serve two main customer segments. One is the many people living in urban areas in the developing world where electric power is delivered via a complex system of highly intertwined and often unrelated wires. These systems mostly have high network losses and many customers are unable to pay according to the power tariffs. The other category is all the people living in off-grid, remote areas. Many of these regions get support today from non-governmental organisations and local entrepreneurs. Utilities ought to spend more effort to bring off-grid electricity to these rural areas. It is apparent that the technical and managerial challenges are quite different for the two main categories of end users.

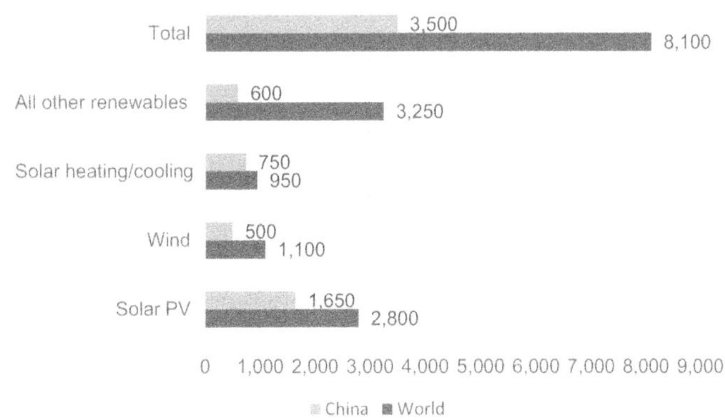

Figure 12.1 Estimated number of direct and indirect jobs in 2017 in the renewable industry (excluding large hydropower) in thousands. The upper bars show that China dominates the job market in the industry. *Source:* Data from REN21 (2017a), Table 1.

Another estimate of job opportunities is presented by World Bank (2017). Many different job skills are required and are categorised in MCI (manufacturing, construction and installations) and O&M (operations and maintenance). Table 12.3 presents a selection of some estimates. Note that the estimates are based on relatively large installations.

12.2.1 Job creation in the solar industry

Almost three million people worldwide were employed (directly and indirectly) by the solar energy sector in 2016 (see Figure 12.1) and the industry continues to expand. As demonstrated in Figure 3.5, most of the solar equipment and the demand are located in Asia, so most of the jobs are there, while in Europe solar PV employment has decreased. Obviously, there is a high demand for jobs not only in manufacturing but also in O&M. In order to expand the use of renewable energy in remote areas it is apparent that not only will more jobs be created but also that the need for more education at all levels is urgent, as noted in Chapter 2 and Chapter 3.8 (Jones *et al.*, 2018). As the share of renewables is set to increase significantly, many more people will have to be employed and educated in the industry compared to today.

Table 12.3 Job opportunities in renewable energy systems.

Energy Source	Construction Time	Construction + Installation *Job Years/MW*	Manufacturing *Job Years/MW*	Operation + Maintenance *Jobs/MW*
	Years			
Wind onshore	2	2.5	6.1	0.2
Solar PV	1	9	11	0.2
Solar thermal	2	5.3	4	0.4
Solar heat		7.4		

Source: Data from World Bank (2017, Figure 4.15).

12.2.2 Job creation in the wind industry

The wind power industry employs more than a million people worldwide, and about half of them are in Asia (Figure 12.1). It is estimated (WEC, 2016, Chapter 10) that wind power alone could support more than four million jobs in 2030, a fourfold increase on today.

Elements of wind power installation require special competences and equipment that are not usually available locally. For example, mounting the turbine and the tower requires a lot more expertise than mounting a solar PV array. Furthermore, mounting tall wind towers may require either high cranes or helicopters. Other activities, such as site planning, can often rely on local labour.

Many parts of a wind turbine, particularly the blades and the nacelle, are large units. It is of course quite an advantage if these parts can be manufactured not too far away. However, this requires available raw material, skilled workers and, of course, a market that makes the manufacturing profitable. To transport large units long distances is costly and not possible in many remote areas.

Some activities require specific expertise and knowledge in the wind sector that are not necessarily available locally at the initial stages of development of the sector. Initial activities in the project planning phase include site selection, technical and financial tasks.

12.3 FINANCING

Financing is a key issue and financing research and development for small-scale operations of both energy generation and water operations needs more attention and funding. There is a risk that interest is focusing on large-scale operations. Lack of investment may be the biggest barrier to universal access to clean water systems powered by renewable energy systems.

Multilateral institutions like the World Bank, the Asian Development Bank (ADB) and the African Development Bank (AfDB) provide important funding to further develop and deploy renewable energy projects. There are also other important institutions like Deutsche Bank's Universal Green Access Program (UGEAP) for Africa. They work with local financial institutions that enable local banks to extend medium- and long-term loans to distributed renewable energy companies and initiatives.

It is still not recognised everywhere that water and energy operations must be handled simultaneously. The delivery chains of water and energy are mostly managed in 'silos', where the silos not only represent professions and sectors but also different institutions. It is apparent that our infrastructures of energy and water should be designed and operated in a more integrated way. This challenge goes all the way up to government agencies and ministries. The Malaysian government may serve as a good role model due to its Ministry of Energy, Green Technology and Water. Collaboration between stakeholders has also to be strengthened. The UN Sustainable Development Goals of clean water and clean energy for everybody are clear and their connection should be better recognised (see 2.1).

12.3.1 Funding in rural areas

Experiences of rural electrification have been reported in, for example, IEA (2013). It is obvious that some kind of subsidy system (where the end users do not pay the real cost of power) is needed in rural areas with low population density and scattered settlements. Since both capital and operating costs must be considered, both these aspects have been reflected in the subsidy systems. However, as noted before, the up-front capital cost is the dominant one in renewable systems. This means that the subsidy system has to change its emphasis towards the dominant capital cost. This is a significant change for the public authorities in charge of subsidising rural electrification. It also brings a change of perspective for the donor agencies.

Compared to traditional diesel-based generators the new technology carries greater technical risks in the sense that there are more components. The system contains not only the electrical devices necessary for solar PV or wind power but also water supply and water treatment components like pumps and desalination membranes. This raises the issue of available spare parts and distribution systems reaching rural areas.

A strong public-private partnership is needed to meet the challenges of new technology and to share the risks. The private sector can provide capacity-building and ensure a quality installation and maintenance of the equipment. The public sector can engage in subsidies and financial support, especially during the initial years, to build up skills and to develop the market. Experiences from successful installations, as documented in World Bank (2017), suggest some important success factors:

- Consideration of the demands, interest and restrictions of local customers, including the desire to pay with mobile payments systems;
- Strong partnerships along the whole supply chain, from the government to private-sector service providers; and
- Adaptation of market dynamics to local conditions to support successful, sustainable clean energy solutions.

12.3.2 Payment models

It is becoming increasingly common to use smartphones to pay for energy services in the developing world. In 2016 there were more than 32 companies in over 30 countries in Africa and developing Asia selling solar PV products to more than 700,000 households. The up-front fee and regular payments are made using mobile money transfers (IRENA, 2017a).

Pay-as-you-go (PAYG) models are becoming common. The key factor is a finance model that matches affordable pricing for the end user with an adequate return on investment for the supplier. For example, PAYG solar companies seek to provide energy services at a price that can compete with consumers' current spending on kerosene, candles, batteries and other low-quality energy services. This of course includes the water system equipment and the potential improvement that it provides, for example by increasing agriculture output using irrigation. Two methods for PAYG are typically offered:

- The end user pays a fee for the *possibility* of owning the system, but never actually owns it;
- The end user will own the system after paying off the system cost. They will make regular payments on a daily, weekly or monthly basis.

Investment in PAYG solar companies has increased from practically zero in 2012 to 223 million USD in 2016 (REN21, 2017).

12.4 FURTHER READING

Hoffman (2017b) presents an insightful analysis and description of the role of subsidies, very much talking from his own experience. The rest of the chapter is an informative description of solar PV financing from a US perspective.

Varadi *et al.* (2018), Chapter 6, is a valuable source of information for anybody who wishes to know more about financing renewable energy in Africa and the Middle East.

Huld *et al.* (2014) have calculated levelised cost of energy (LCOE) for PV systems in Africa and the Middle East, both grid-connected and off-grid PV systems.

IRENA (2017c) presents a lot of facts concerning jobs and renewable energy.

IEA (2016b) discusses the system value of solar PV and wind. The report emphasises the importance to maximise overall value to the power and wider energy system rather than minimising the generation cost of wind and solar power in isolation.

Chapter 13

Land use for energy

> "Changes in land use associated with energy development can have a significant impact on the quality of the physical, social, economic, and visual environment. Local air quality, water quality, water availability, noise levels, the municipal tax base, land values, job opportunities, and the character of the community itself may be affected."

US Department of Energy Report about land use and energy in 1975.

Land use is a crucial issue when addressing the energy-water-food nexus. The competition for land is apparent in several ways. The world is facing increasing food demands due to both increasing average incomes and the rise in population. At the same time, it is feared that the agriculture yield will drop because of water scarcity.

It is argued that solar PV and wind power may require fertile land and consequently threaten food production. On the contrary, solar PV provides a large benefit by using not land but spare rooftops. The land area requirements for different types of electricity generation can be compared (Olsson, 2015). Large hydropower requires a certain area for the reservoir. Often there is a multi-purpose reservoir, used for water storage and flood protection besides hydropower. For wind power the total area enclosed by the site boundary is defined as the

© IWA Publishing 2018. Clean Water Using Solar and Wind: Outside the Power Grid
Author: Gustaf Olsson
doi: 10.2166/9781780409443_0159

land use, although the area between the towers can often still be utilised for agriculture or forest. Offshore wind will of course have an environmental impact as well, but the area seldom competes with other uses. Solar PV does not necessarily need to occupy fertile land. Small-scale PV and solar heating installations have minimal land impact, as they are actively integrated into buildings and structures they serve.

The power and energy outputs from a given area are summarised in Table 13.1. It is obvious that solar PV is very competitive in regard to land use, even if the capacity factor is relatively low for the actual area. In the US the National Renewable Energy Laboratory (NREL) (www. solarindustrymag.com/online/issues/) has studied the land footprint of utility-scale solar generation. There is a wide range of total land use. The average total land use was estimated at 8.9 acres (around 36,000 m^2) per MW, or around 28 MW/km^2 (compare with Table 13.1).

Table 13.1 Energy output from a given area for different renewable sources (Olsson, 2015).

	Hydropower	Wind	Solar PV
Power density MW/km^2	0.1–17	5–8	20–110
Capacity factor	0.6	0.3	0.2
Annual energy output GWh/km^2	0.5–90	13–21	35–190

The wind power land use requirement depends on both the size of the turbines and the extent of the terrain. In a hilly area the wind turbines may be located along the ridgelines. Wind towers on flat terrain are often positioned more uniformly and may require a larger land area. The footprints for wind farms in the United States average around 333,000 m^2 per MW or a power density of 3 MW/km^2 (NREL, 2012b).

The space between the wind turbines on a farm can be large. However, only a small fraction (3–5%) of the land required is disturbed by the wind energy structures. The rest of the land could be utilised for agriculture and transport links. Naturally, in remote areas there will seldom be large wind farms; instead there will be stand-alone wind turbines, with quite a small land footprint.

The global share of rooftop PV systems is not known. In many countries there are no separate statistics for rooftop PV and utility-scale PV. World Energy Council (WEC, 2016, Table 7) reports that in only four major solar PV countries (Germany, Japan, the US and Australia) do the land savings as a result of rooftop installations exceed 200,000 acres or 85,000 hectares (=850 km^2).

Solar panels are used in innovative ways to save both land and water (see 8.3.4; WEC, 2016). In Japan the 13.7 MW Yamakura floating solar power station is composed of more than 50,000 solar modules, covering a water surface area of 180,000 m^2 : or 76 MW/km^2. They are mounted on the Yamakura Dam reservoir, located in the Chiba Prefecture east of Tokyo. The panels will reduce water evaporation from the dam as well as saving fertile land. The plant was put into full operation in March 2018.

A similar structure is being developed in India . In the first stage of a solar panel project in the province of Gujarat in north-western India, a 750 m section of a water canal is covered with solar panels, generating 1 MW of electric power. Covering the canal with solar panels will save agricultural land as well as decreasing the water loss via evaporation (Shukla *et al.*, 2016). According to WEC (2016) the solar panels of this canal could save a lot of land, five acres per MW. This corresponds to a solar panel power of 50 MW/km^2 (compare Chapter 8.2–8.3).

Chapter 14

Water operations using renewables – some cases

"By seeking and blundering we learn."

Johann Wolfgang von Goethe 1749–1832.

Various renewable energy applications for water operations are described here. Some of them are utility-scale and others are small-scale. Our purpose is to show both the applicability at various scales and, when relevant information is available, the cost/efficiency patterns. Some of the installations are connected to electric power grids that can be used for balancing the production and load and others are isolated and off-grid.

14.1 DEVELOPING COUNTRIES VERSUS HIGH-INCOME COUNTRIES

The need to balance production and load can look quite different in high-income and in low-income regions. People in higher-income areas will require power availability round the clock and will usually not accept too many interruptions in power delivery, whereas people getting their first electric power source will probably be more tolerant of lack of delivery during the night. This will define the ambition of the storage capacity (see Chapter 10.2).

Economy is crucial. As noted by Professor Akhlesh Lakhtakia, Penn State University, U.S.: "Poor people don't need the most efficient sources. They need affordable ones, and a helpful nudge to improve their lives; that motivated us in our research." Research is ongoing into developing solar cells that are less efficient (up to 17%) than most on the market but can produce a viable level of electricity at a greatly reduced production cost. Instead of using silicon the researchers are exploring indium gallium nitride, which could give some advantage with its semiconducting properties.

It is apparent that renewable energy can play a major role in extending energy access to communities in the developing world. However, many of these countries suffer from a lack of technical expertise to implement these facilities. As noted previously, a shortage of local human resources is a key barrier to fulfilling the high potential of renewable development. Therefore, it is important to ensure technical education in these regions. High-income countries have a huge responsibility to make this happen. This can enable the development of local industries to provide the country with renewables.

14.2 IRRIGATION AND WATER PUMPING

Irrigation water requirement (IWR) is site-specific and depends on the particular crop. IWR values can be anywhere between 20 and 70 m^3/hectare/day (Campana *et al.*, 2015). As a comparison, a precipitation of 1 *mm* corresponds to 10 m^3/hectare.

Example 14.1: *Pumping for Irrigation Using Solar PV, Senegal.*

A low-cost and simple solar pumping system was implemented in Senegal in 2013 (www.youtube.com/watch?v=bPvPJuvLw9Q). The potential source of water is a small river nearby and the pumping system is used for irrigation. Solar panels were becoming affordable and the challenge was to find other affordable components of the system. There are reliable solar pumps available, but they are more expensive than the solar panels.

The system has five solar panels with a capacity of 5 · 80 $W = 400$ W. In Senegal solar panels are easily available even at the roadside.

The cost of these panels was less than 1 USD/*W*. The panels are mounted on a wheel cart that can easily be pushed around. This innovative thinking made it possible to turn the panels into the best possible position, towards the sun and avoiding shadows from trees: in other words, they became a manual tracking system. The panels were placed at the irrigation site at some distance from where people live and could easily be stolen at night. With the cart the panels can instead be moved to a safe stored location overnight.

The aim was to find an affordable pump and keep the cost down by avoiding using batteries. By using a DC pump the cost of DC/AC conversion was avoided. The drawback is that DC motors have a shorter life than AC motors. Marine pumps were found to be a good choice, but many of them are designed only for low heads (see Chapter 4). The actual pump can deliver around 15 m^3/hour or 4 litres/second at zero head. It can work at 5–6 m head, but then the flow rate is lower (compare Figure 4.2) albeit sufficient for the purpose. In this case the motor is assumed to have a life of around one year. However, the profit from the irrigation could pay for the replacement of the motor. Altogether the pumping system cost was less than 1,000 USD. No electronic controllers were used, only a simple circuit breaker. The pump can irrigate around half a hectare (=5000 m^2). A flow rate of 4 m^3/hour for six hours will provide almost 50 m^3/hectare/day.

14.3 DESALINATION

The integration of renewable energy resources in desalination and water purification is becoming more viable as costs of conventional systems increase, commitments to reducing greenhouse gas emissions are implemented and targets for exploiting renewable energy are set. Many PV-based desalination systems have been demonstrated throughout the world, especially in remote areas and on islands. The examples and cases illustrate both utility-scale installations and small-scale implementations. They are shown to exemplify both efficiency and economy.

14.3.1 Solar PV desalination installations

Example 14.2: *Village of Ksar Ghilène, Tunisia*

A solar PV-based system for water supply was already in existence in 2005 in the village of Ksar Ghilène, located in southern Tunisia (www.

adu-res.org/pdf/ITC.pdf). The village has about 300 inhabitants, 50-plus families, who live off agriculture and cattle raising. The nearest drinking well is located 60 km away. The solution to the water supply challenge was a reverse osmosis (RO) plant supplied by a solar PV system. The project was financed internationally, and technical support provided by the Canary Island Institute of Technology, which had experience of supplying drinking water through stand-alone systems.

The village's daily water consumption during the summer is about 15 m^3, the solar irradiation annual average 5,600 kWh/m^2 (compare Table 8.1) and the annual average temperature 26°C, varying from 0 to 45°C.

The raw water is pumped from an artesian brackish (around 3,500 ppm) water well located inside the oasis, some 2 km from the village. The water is desalinated in an RO unit. The water is used for irrigation of palms and crops and for tourist services.

The electric energy is used for the feedwater pump from the artesian well and for the compressor in the RO unit. Some of the system parameters are:

PV generator	Peak power 10 kW_p; operation time = eight hours/day
Feed pump	Max power = 1 kW; pumps water from the well through a 2 km pipe; flow rate = 3 m^3/hour; pressure = 3 bar
High-pressure pump for the RO unit	Requires 3 kW for pressure <15 bar; flow rate = 3 m^3/hour
Disinfection	200 W
Lighting	250 W
Battery storage	capacity = 600 Ah; C10 batteries ($\rightarrow$ discharge current will discharge the battery in ten hours)
Produced water	Production = 15 m^3/day; salinity <500 ppm
Specific power	2 kWh/m^3 of produced water

Example 14.3: Abu Dhabi

The United Arab Emirates (UAE) is considered a "water-scarce" country. It has just 83 m^3 of water per person per year – well below the UN scarcity threshold of 1,000 m^3. As a consequence, UAE relies to a large extent on seawater desalination to satisfy the demand for water supply.

Mascara Renewable Water has developed an off-grid, solar-powered desalination solution in Abu Dhabi (Masdar, 2018). The plant uses a beach well to obtain seawater from a borehole near the sea. The natural sand filtration of the beach well eliminates the need for a dedicated pre-treatment system. The intermittent power production is compensated for by a hydraulic energy accumulator used as storage. The system is powered by a 30 kW_p PV plant, and the system operates only during sunlight hours, producing 30 m^3/day. Biofouling is avoided by automatically flushing the membranes before sunset every day.

A number of identical desalination plants have been designed, based on solar energy and located in isolated desert areas of Abu Dhabi, outside the power grid. A typical solar system is built up of 300 m^2 panels that will produce a maximum of 45 kW, in other words 150 W/m^2, which is in the same order of magnitude as described in Chapter 8.

The first installation was completed in 2009. The desalination plants designed by Hitachi are pumping saline groundwater and applying reverse osmosis to clean the water. The salinity ranges from brackish water to 35,000 mg/l, similar to seawater. The production of the system is 4 m^3/hour of fresh water.

The groundwater is first pumped to a storage tank before treatment in the RO unit. Even in a sunny area like the Abu Dhabi desert the sunlight may be shaded during the day, for example due to sandstorms. Therefore, a battery backup is provided.

There is an evaporation pond, designed to get rid of the brine reject (see 5.3.4).

Example 14.4: Gran Canaria, Spain

A solar PV-powered system for the desalination of seawater, called DESSOL, has been installed close to the beach on Gran Canaria (Espino *et al.*, 2003). The desalination system is based on RO and produces an annual average flow of 3 m^3/day (or 0.4 m^3/hour) during eight hours of operation in the summer and six hours in the winter. Some of the system characteristics are:

Solar panels	Total capacity = 4.8 kW_p 64 modules of 75 W_p each
Battery storage	19 kWh
Well pump	Max altitude above the sea = 3 m Capacity = 1 kW Max flow rate = 2.5 m^3/hour; pressure = 3 bar
High-pressure pump for RO	Capacity = 2.2 kW Max flow rate = 1.5 m^3/hour; pressure = 60 bar
Effluent salinity	<500 mg/l

The authors describe an interesting aspect of the operation to protect the membranes. When the plant is shut down at the end of every day there is a flushing process, using produced water, to keep the membranes clean. The membranes are submerged in low-salinity water, preventing deterioration due to their intermittent operation (see 10.1.3; Lienhard *et al.*, 2016).

Example 14.5: *Solar Heating for Desalination, California*

In the California Central Valley, the Panoche Water District is using a solar thermal system for desalination (Lavelle, 2015). The solar energy is not used to produce electricity. Instead, parabolic trough mirrors turn the solar radiation directly into heat to distil salty water.

Example 14.6: *India*

India has a highly seasonal pattern of rainfall, with 50% of precipitation falling in just two weeks. The Central Water Commission estimates that the total annual rainfall in the country is 4,000 billion (4 · 10^{12}) m^3. The utilisable or internally renewable water resources are estimated to be 1,200 billion m^3. The annual water demand is increasing: it was around 800 billion m^3 in 2010 and is estimated to approach 1,500 billion m^3 in 2050. This will not be sustainable, and many regions will face severe water shortage. Furthermore, the impact of climate change will lead to variation in rainfall patterns and evaporation rates. A growing population and industrialisation will put a lot of pressure on water resources. Several regions are already suffering from excess contaminating factors like salinity, fluoride, iron, arsenic, heavy metals and microbial contaminations of groundwater. The need for sustainable water-supply solutions is urgent.

Example 14.7: *Village Installation, Rajasthan, India*

In Kotri, a small village of 300 families in the region of Rajasthan in north-western India, a solar-based RO plant has been put into operation. The plant produces drinking water for more than 1,000 residents from both Kotri and surrounding villages (IRENA, 2015b). Brackish water from a nearby lake is pumped through the RO plant and produces around 600 litres/hour of water for six hours every day. The salinity of the water is reduced sufficiently to make the water drinkable. The RO plant is served by a 2.5 *kW* power plant. The village is in fact connected to the grid, but the supply is very unreliable with only three hours/day of power most of the time. The solar-powered system guarantees six hours of electric power supply, which gives some surplus power for light, fans and computers.

Example 14.8: *Village Installation, Andhra Pradesh, India*

A rural village in India gives a typical example of the application of solar PV to treat water (WEC, 2016, Chapter 8). The SANA organisation (Social Awareness Newer Alternatives) identified a village that had no access to clean drinking water and where the power supply was irregular: the N. Chamavaram village in the state of Andhra Pradesh in south-east India. Energy from the solar PV system has been used to purify contaminated water to WHO drinking-water standard. This is a typical example of decentralised water supply where the raw water intake can be either contaminated well water or used water that is reused. The capacity of this system is 1,800 m^3 of water yearly or 5 m^3/day. This will supply 1,000 schoolchildren from economically backward homes with five litres of water daily for their families, who live in slums nearby.

14.3.2 Wind power desalination installations

The electrical and mechanical power generated by a wind turbine can be used to power desalination plants, in particular RO units. In general, wind power-based desalination can be one of the most successful options for seawater desalination, especially in coastal areas with high wind potential. As for solar PV, wind desalination has the drawback of the intermittence of the energy source. Possible combinations with other renewable energy sources, batteries or other energy storage systems can provide smoother operating conditions. As with solar PV, water desalination itself can provide an excellent storage opportunity in the case of electricity generation exceeding demand.

Various wind-based desalination plants have been installed around the world, including in Gran Canaria, Canary Islands (wind-powered RO, seawater, 5–50 m^3/day), Fuerteventura, Spain (wind-diesel hybrid system, seawater, 56 m^3/day) and the Centre for Renewable Energy Systems Technology in the United Kingdom (wind-powered RO, seawater, 12 m^3/day) (Kalogirou, 2005; Gude *et al.*, 2010; Al-Karaghouli and Kazmerski, 2011).

Example 14.9: Sydney, Australia

Sydney Water desalination plant supplies about 15% of the water for Australia's most populous city (www.metrowater.nsw.gov.au). To help minimise the carbon footprint of the desalination plant, the power requirements are being 100% offset with renewable energy generated at a 67-turbine wind farm near Bungendore, about 270 *km* to the south. The wind farm generates more than enough electricity to power the plant: the plant needs around 42 *MW* while the wind farm's capacity is 132 *MW*. Notice that the 132 *MW* is a peak capacity, so the desalination plant needs some 32% of the wind power peak capacity, which is close to the wind power efficiency.

Example 14.10: Perth, Australia

The city of Perth in Western Australia has a large desalination plant, the Perth Seawater Desalination Plant, producing around 140,000 m^3/day (www.watercorporation.com.au). With energy demand at 3.5 kWh/m^3, this will require around 490 *MWh*/day, which corresponds to a continuous power supply of 20 *MW*. The power is delivered from a wind farm located 260 *km* away from the plant. It is documented that the desalination plant requires an average wind power peak capacity of 82 *MW*, which means that the average delivery of power is about 25% of the capacity.

Example 14.11: *Texas, U.S.*

Another example of wind-powered desalination is from Texas, U.S. (Swift *et al.*, 2009) in a region suffering from severe water scarcity and depending on deep high-salinity aquifers. RO treatment of this kind of brackish water is a realistic and economically feasible solution. There is not only high salinity (around 2600 *mg/l* of total dissolved solids); arsenic and fluoride concentrations are also high. RO technology would lower these to acceptable limits. A feasibility study has been made for a municipal, integrated wind-water desalination system for an inland small community. The study from Texas demonstrates that the integration of the two relatively mature technologies of wind energy and RO becomes an attractive option for addressing an emerging threat to any region heavily dependent on affordable energy and potable water. In the Texas study a small 5 *kW* wind turbine provided the energy for an RO desalination plant with the capacity of about 6 m^3/day. The energy requirement was found to be around 0.82 *kWh/m^3* of treated water. The Swift report gives a detailed account of not only wind power supply but also operating and maintenance experiences of desalination facilities.

14.4 FURTHER READING ON DESALINATION AND RENEWABLE ENERGY

There is a lot of literature and information concerning desalination using renewable energy. Some useful sources are:

* *Desalination* journal (Elsevier);
* Elemental Water Makers, a Dutch start-up company offering desalination using renewable energy (www.elementalwater makers.com);
* Fraunhofer Institute for Solar Energy Systems ISE (www.ise. fraunhofer.de);
* Lenntech, a company offering solutions for desalination systems (www.lenntech.com);
* EIP Water (The European Innovation Partnership on Water): an initiative within the EU 2020 Innovation Union. The EIP Water facilitates the development of innovative solutions to address major European and global water challenges (www.eip-water.eu).

Part V
The Future

We have mentioned the future in other parts of the book. It is always risky to predict the unpredictable, but still we dare to do so, not only to have something to laugh at in a few years but also to express an ambition for today. For example, before 2010 there was quite a slow increase in solar PV installations, and I was fairly hesitant about its potential to replace conventional energy sources in the next few decades. However, its development took off soon after that at a rate that we had thought impossible. Technical, commercial and environmental forces have cooperated to make it possible. In the last chapter of the book we dare to express some ambitions for the next decades.

Chapter 15

Outlook to 2030 and further

"World electricity demand is expected to grow by more than 50% by 2030,
mostly in developing and emerging economies.
To meet this demand while also realising global development
and sustainability goals, governments must implement
policies that enable solar to achieve its full potential."

Adnan Z. Amin, IRENA Director-General, 2016.

The year 2030 is only 12 years away. Many organisations have published predictions for 2030 and further ahead and they are in general ambitious and optimistic. Predicting the unpredictable may give us a laugh when we read it again after a few years; but looking back is easy. It can be useful, but speculation about the future propels us forward; and there is no time to waste.

15.1 PREDICTIONS FOR RENEWABLES

IRENA (2016b, 2017a) envision an optimistic and ambitious goal for renewables. According to their prediction and ambition there will by 2030 be a doubling of the energy share from renewables, up to 36%.

The share of solar PV is expected to increase six times, up to 7% of the global power generation in 2030. Some even more optimistic predictions say 13%. Naturally, this will require a combination of technology development, policy development and active financing instruments and investments.

The ownership of power production is going through a crucial change. Electric power systems were once dominated by monolithic state agencies and large corporations. With small-scale power generation there is now an increasing number of owners and producers. This will also change attitudes and responsibilities for the systems.

Individual welfare issues like personal health and education as well as climate benefits have been emphasised and should not be underestimated. Solar PV and wind have limited water requirements; instead they are power sources for water supply and water reuse.

It is apparent that prices for both power generation and energy storage will fall rapidly. However, cost and availability of capital will still be a major challenge since most of the price tag for renewables is up-front capital cost.

There are obvious reasons to look carefully at the potential development of sub-Saharan regions. IRENA (2013) is an informative account of the development of renewables in Africa. IEA has produced *African Energy Outlook,* the first of its kind to provide a far-reaching picture of the energy situation in sub-Saharan countries today and in the future (IEA, 2017d). This region contains 13% of the global population, but only 4% of its energy demand. However, since 2000 energy use in the region has risen by 45%, which is one indicator of rapid economic growth.

As noted in Section 1.3, grid connections are often unreliable. This makes it necessary to invest in costly private use of backup generators running on diesel or gasoline. IEA predicts that 70% of those gaining access to electricity in rural areas by 2040 will be connected to mini-grids or off-grid systems. A gloomy prediction, however, is that more than 500 million people, mainly in rural areas, will still be without electricity in 2040. IEA predicts that around two-thirds of the off-grid and mini-grid rural systems in 2040 will be powered by solar PV, wind or hydropower. Renewable systems will be increasingly competitive compared to diesel generators.

15.2 DESALINATION RESEARCH AND DEVELOPMENT

It is an important research topic to adapt desalination technology to the variability of renewable energy sources. The intermittent power production of solar PV implies that solar-powered desalination membranes will undoubtedly operate outside their optimum operation window. Therefore, it is essential to investigate the long-term reliability of variable and intermittent operation (non-stable pressure and flow rate) on membrane reliability. This includes salt passage at the ion level, fouling (colloidal, biological and organic) and scaling (mineral and salt precipitates) of membranes (Lienhard *et al.*, 2016; Zaragoza, 2018).

A lot of research is ongoing into finding new types of membranes for RO. Graphene is one of the strongest materials known. A graphene membrane is only one atom layer thick and has much higher permeability than today's membranes. This means that less pressure and electric energy will be required for desalination.

Other challenges include the overall design and optimisation of the components in the integrated system. Control and management aspects should be taken into consideration at the design stage so that the energy flow in the system can be assured for all possible operational conditions.

Another challenge is related to small-scale technology. Small-scale operations struggle to be cost-competitive with utility-scale water operations.

15.3 SOFT ISSUES

Supplying clean water using renewable energy includes many "soft" issues, as described in the preceding chapters.

Integration between energy supply and water operations should be better understood. It is essential that energy production is adapted to the needs from the loads such as the pumps, desalination units and water remediation units.

15.3.1 Education and training

Off-grid energy generation and water operations are small-scale and decentralised. There are lessons to learn from experiences so far, and some are summarised in IEA (2014). There are problems beyond technology challenges. Many projects have failed due to inappropriate or unclear organisation of operating and maintaining systems. There must be a good understanding of the needs of the users. Similarly, it is important to make realistic estimates of current and future power needs. There are two aspects of education and training: (1) the required education to make the best use of renewable energy and water facilities and (2) the impact of access to electricity on education.

Developments of projects occur locally. We have emphasised the urgent need for training and education. There is a need for increased knowledge at all levels: among the customers and users as well as locals who can be part of the workforce to mount, install and maintain clean water operations in the neighbourhood. Such developments can create local jobs and incomes.

Access to water and energy will of course develop rural areas in terms of agricultural improvements, productivity, income and better living conditions.

Healthcare, education, home environment: all benefit from access to modern energy and to clean water. The modularity of the energy and water operations also means that they can be customised to individual needs and applications.

The impact of access to electricity on education is illustrated in 2.2. Available electricity can also contribute to people's daily habits and activity scheduling, and people's active day may be extended. These issues are analysed in detail in Riva *et al.* (2018).

15.4 FURTHER READING

SNV, a not-for-profit international development organisation, was founded in the Netherlands more than 50 years ago. SNV has a long-term, local presence in the poorest countries in Asia, Africa and Latin America. It has published an excellent training manual for solar PV (SNV, 2015).

IEA (www.iea.org), IRENA (www.irena.org) and the World Bank (www.worldbank.org) make regular updates to their predictions for renewable energy. Water and sanitation aspects are the focus for several speciality groups within IWA (www.iwa-network.org).

Appendix 1

Glossary

Activated sludge process: a biological wastewater treatment by which bacteria that feed on organic waste are continuously circulated and put in contact with organic waste in the presence of oxygen to increase the rate of decomposition.

Aerobic: "with oxygen", used for biological treatment systems characterised by the presence of oxygen, mostly as oxygen dissolved in water.

Alternating current (AC): electric current that reverses direction 50 or 60 times per second.

Anaerobic: requiring absence of air.

Aquifer: large body of permeable or porous material situated below the water table that contains or transmits groundwater.

Battery: a type of battery that can be given a new charge by passing an electric current through it is also called a storage battery. A lead-acid battery uses plates made of pure lead or lead oxide for the electrodes and sulphuric acid for the electrolyte; these remain common for off-grid installations. A lithium battery uses a liquid lithium-based material for one of its electrodes. A flow battery uses two chemical components dissolved in liquids contained within the system and most commonly separated by a membrane.

Blackwater: water from toilets. Compare greywater.

Brackish water: water that is neither fresh nor salt.

Brine: a solution containing large concentrations (higher than seawater) of sodium chloride and other salts.

Capacity: the rated capacity (for example 1 MW) of a power-generating plant, referring to the instantaneous electricity output.

Capacity factor: the ratio between the actual output of a power plant and the theoretical output of the same plant operating at full capacity. The capacity factor is considered over a specific time period, for example a year.

COD: chemical oxygen demand. Method of measuring the content of all oxidable substances in the water.

Conversion efficiency: the ratio between the produced energy from an energy conversion device and the energy input into it. For a solar PV the conversion efficiency gauges the percentage of solar (light) power reaching a module that is converted into electric power. If 100 kWh of solar radiation is received and 15 kWh electricity is generated, then the conversion efficiency is 15%.

Cut-off voltage (in a battery): the minimum allowable voltage. It is this voltage that generally defines the "empty" state of the battery.

Direct current (DC): the unidirectional flow of electric charge. A battery is a good example of a DC power supply. The electric current flows in a constant direction, distinguishing it from alternating current (AC).

Depth of discharge (DoD): the percentage of battery capacity that has been discharged expressed as a percentage of maximum capacity.

Desalination: reducing the contents of total dissolved solids or salt and minerals in sea- or brackish water into fresh water.

Discharge time: defined as the energy capacity divided by the nominal power. The time over which the energy stored in a storage device can be discharged at the nominal power rating.

Distributed generation: a small-scale power generation technology that provides electric power at a site closer to customers than central power plant generation.

Distributed renewable energy: energy system where generation and distribution occur independently from a centralised network. The system is close to the point of consumption.

Efficiency: the ratio obtained by dividing the actual power or energy by the theoretical power or energy.

Energy capacity: the amount of energy that can be stored and recovered from a storage device, expressed in joules or *kWh*.

Fossil fuel: fuels such as coal, crude oil or natural gas, formed from remains of plants and animals.

Fouling: the process of becoming dusty or clogged, for example, in which undesirable foreign matter accumulates in a bed of filter or ion exchanger media, clogging pores and coating surfaces, thus inhibiting or delaying proper bed operation. The fouling of a heat-exchanger consists of the accumulation of dirt or other materials on its wall, causing corrosion and roughness and ultimately leading to a lowered rate of efficiency.

Fresh water: water with less than 1,000–2,000 parts per million (*ppm*) of dissolved salts.

Generator: device that converts the rotational energy from a turbine to electric energy.

Greywater: domestic used water from kitchen, bathroom and laundry sinks, tubs and washers. Compare blackwater.

Groundwater: water that is below the land surface in pores or crevices of soil, sand and rock, contained in an aquifer. If the groundwater has a negligible rate of natural recharge on a human timescale it is often called fossil or non-renewable water.

Hydropower: the harnessing of flowing water – using a dam or other type of diversion structure – to create energy that can be captured via a turbine to generate electricity. Large hydropower is typically a power rating of more than 30 *MW*. Small hydropower is typically less than 10 *MW*.

Intermittent electricity: electric energy that is not continuously available due to external factors that cannot be controlled. Sources of intermittent electricity include solar and wind power. Their electrical output cannot be used at any guaranteed time to meet fluctuating electricity demands.

Inverter (solar): a power electronics device that converts power from solar PV modules or batteries in DC form into alternating form (AC) at a required frequency and voltage output. This can be used by a local, off-grid network. The inverter circuit's AC output voltage waveform is not a sine wave but usually a square wave or a distorted sine wave.

Levelised cost of energy/electricity (LCOE): measure of the total cost (for example, measured in cost per *kWh*) to produce electricity, including capital cost, operating, maintenance and fuel costs. The cost is discounted back to a common year using a discount rate.

Levelised cost of storage (LCOS): the average cost to store and discharge energy (cost per *kWh*). The LCOS is calculated over the entire lifetime of the storage and includes the capital and operational costs. The cost is discounted back to a common year using a discount rate.

Off-grid renewable energy: renewable energy generation that is not connected to a larger electricity system or network.

Osmosis: the spontaneous net movement of solvent molecules (such as water molecules) through a semi-permeable membrane into a region of higher solute concentration (such as seawater), in the direction that tends to equalise the solute concentrations on the two sides.

Photovoltaic: production of electric current at the junction of two substances exposed to light. A photovoltaic cell (PV cell) is a specialised semiconductor diode that converts visible light into direct current (DC). Some PV cells can also convert infrared (IR) or ultraviolet (UV) radiation into DC electricity.

Reverse osmosis: type of membrane filtration (see osmosis).

Scaling: precipitation of solid substances on the membrane in nano- and reverse osmosis filtration.

Silicon: the basic material used to make solar cells. It is the second most abundant element in the earth's crust, after oxygen. Silicon is a metal and, therefore, its atoms are organised into a crystalline structure.

Solar home system (SHS): a stand-alone (not connected to the electricity grid) system composed of a relatively low-power photovoltaic module, a battery and sometimes a charge controller, which can power small electric devices and provide modest amounts of electricity to homes for lighting and radios.

State of charge (SoC): the present battery capacity as a percentage of maximum capacity. SoC can be calculated by integrating the current over time.

Surface water: water pumped from sources open to the atmosphere, such as rivers, lakes and reservoirs.

System cost: this includes all components of a renewable energy system other than the photovoltaic panels or the wind turbine. This contains wiring, switches, a mounting system, one or many inverters, a battery bank and battery charger. Other soft costs include: financing, mechanical installation, electrical installation, system design, customer acquisition, permitting, inspection/certification, connection, operation and maintenance.

Variable renewable energy: see Intermittent electricity.

Water consumption: the volume withdrawn that is not returned to the source (i.e. it is evaporated or transported to another location) and so by definition is no longer available for other uses (see Water withdrawal).

Water footprint: the amount of fresh water utilised, for example for energy production (such as litres/*kWh*).

Water stress: defined as when renewable annual fresh-water supplies fall below 1,700 m^3 per person and year; *water scarcity* (limitation to economic development and human health and well-being) is below

1,000 m^3 per person; and *absolute scarcity* (main constraint to life) below 500 m^3 per person.

Water treatment: process of removing contaminants from water or used water in order to bring it up to water quality standards and for storage in fresh-water reservoirs.

Water withdrawal: the volume of water removed from a source; by definition withdrawals are always greater than or equal to consumption (see Water consumption).

Wind park (wind farm): a group of wind turbines in the same location used to produce electricity.

Appendix 2

Conversion of units

A2.1 POWER AND ENERGY

It is important to distinguish between power and energy. Power is energy *per time unit*, the *rate* of energy production or consumption. The SI (International System of Units) or metric unit of energy is joule (J) and 1 J is defined as 1 Ws (wattsecond).

1 J is the designated name for the work 1 *newton · metre*, in other words, the force 1 *newton* along the length 1 *metre*. The basic *power* unit watt (W) is defined as 1 J/s.

1 J = 1 Ws (wattsecond)
1 megajoule (MJ) = 10^6 J
1 gigajoule (GJ) = 10^9 J

Kilowatt-hour (kWh) is a standard unit of electric energy. Since 1 kW (kilowatt) = 1,000 W and 1 hour = 3,600 seconds we get:

1 kWh = (10^3 W) · (3600 s) = 3.6 · 10^6 Ws = 3.6 · 10^6 J = 3.6 MJ (exact).

1 MW (megawatt) = 10^3 kW = 10^6 W (typically, a large industrial plant or wastewater treatment system has a power rating of the order MW).

© IWA Publishing 2018. Clean Water Using Solar and Wind: Outside the Power Grid
Author: Gustaf Olsson
doi: 10.2166/9781780409443_0185

In a thermal power plant, we must distinguish between the *electric power* (MW_e) and the *thermal* power (MW_{th}).

1 GW (gigawatt) $= 10^3$ MW
(a typical power capacity of a large nuclear power plant).
1 $TWh = 1,000$ $GWh = 10^6$ $MWh = 10^9$ $kWh = 10^{12}$ Wh

The annual electric energy use for a nation is typically expressed in *TWh*. For example, all used water treatment in Sweden requires annually about 0.6 $TWh = 600$ GWh. Consequently, there is an average power level of $600/8,760 = 0,068$ $GW = 68$ MW every hour of the day and night. With nine million inhabitants, every citizen uses on average 7.5 W for used water treatment. About the same power and energy is used for supplying drinking water.

We still see the old unit *horsepower* in American publications:

1 horsepower $= 1$ $hp = 746$ W

A2.2 PRESSURE

The metric unit for pressure is *pascal* (*Pa*), where 1 $Pa = 1$ Newton/m^2, which is a very low pressure.

1 bar $= 10^5$ $Pa = 0.1$ MPa;   1 $MPa = 10$ bar

Old units are:

1 psi (pound/inch^2) $= 6,895$ Pa;   1 bar $= 14.5$ psi

A2.3 HEAT CONTENT

Before it was realised that heat was a form of energy, it was measured in terms of its ability to raise the temperature of water. The calorie and the British thermal units were defined in this way.

Calorie (cal): In a traditional definition one calorie is the amount of heat required to raise the temperature of 1 gram of water by 1°C, from 14.5°C to 15.5°C.

British thermal unit (Btu) is the English system analogue of the calorie.

1 *Btu* is the amount of heat required to increase the temperature of one pound of water (which weighs exactly 16 ounces) by 1°F.

1 $Btu = 251.9958$ $cal.$

In 1948 it was decided that, since heat is a form of energy, the SI unit for heat should be the same as for all other forms of energy, the joule. One *cal* is defined to be 4.1860 *J* (exactly) with no reference to heating of water. (The "calorie" used in nutrition is really a kilocalorie.)

The relationship between the *kWh* and the *Btu* depends upon which "Btu" is used.

1 megajoule (MJ) = 10^6 J = 0.278 *kWh* = 947.8 *Btu*; 1 *kWh* = 3412 *Btu*
1,000 *Btu* = 0.293 *kWh*;   100,000 *Btu* = 1 therm

The unit "quad" is often used in the U.S.:

1 quad = 1 quadrillion (10^{15}) *Btu* = 1.05506 * 10^{12} megajoule (MJ) =
1.055 *EJ* (note that quadrillion in Europe = 10^{24})

A2.4 VOLUME, AREA AND LENGTH

Some common metric length units:

1 micron = 1 micrometre = 10^{-6} *m*
1 angstrom (Å) = 10^{-10} *m* (named after the Swedish physicist
A. J. Ångström, 1814–1874)
10 Å = 1 *nm* = 10^{-9} *m*

Metric area units:

1 *hectare* = 100^2 m^2
1 km^2 = 1000^2 m^2

Non-metric units:

1 US gallon = 3.78 litres; 1 UK gallon = 4.546 litres = 1.2 US liquid
gallons
1 American barrel = a liquid measure of oil, usually crude oil = 42
US gallons = 159 litres
Barrel of oil equivalent refers to the energy equal to a barrel of
crude oil,

= 5.8*10^6 *Btu* or 6119 *MJ*

Acre-foot (the volume of one acre (4,047 m^2 or 43,560 ft^2) with
the depth of 1 foot (0.305 *m*)) is often used, particularly in the

U.S., to denote the annual water consumption for a family or for irrigation.

1 acre-foot=$4,047\,m^2 \cdot 0.305\,m$=$1,233.5\,m^3$ (=$43560\,ft^3$=326,700 gallons).
1 cubic foot = $0.305^3\,m^3$ = $0.0284\,m^3$ = 28.4 litres; $1\,m^3$ = 35.25 cubic feet

A2.5 MASS

1 pound (lb) = $0.4536\,kg$
1 metric ton = 0.984 long ton or English ton

A2.6 CONCENTRATION

Concentrations are often measured in mg/l (=ppm, parts per million) = kg/m^3

A2.7 WATER USE IN ENERGY PRODUCTION/ GENERATION

In some US sources we find $gallons/MBtu$ (millions of Btu):

1 $MBtu$ = 293 kWh = 1054 MJ
1,000 gallon/$MBtu$ = 12.9 litres/kWh = 3.59 $litres/MJ$
1 $litre/MJ$ = 279 gallons/$MBtu$

A2.8 ENERGY USE IN WATER OPERATIONS

kWh/million gallons:

1,000 kWh/million gallons = 1 MWh/million gallons = 0.264 kWh/m^3
1 kWh/m^3 = 3,780 kWh/million gallons = 3.78 MWh/million gallons

kWh/acre-foot:

1,000 kWh/acre-foot = 1 MWh/acre-foot = 0.81 kWh/m^3
1 kWh/m^3 = 1230 kWh/acre-foot = 1.23 MWh/acre-foot

Bibliography

ABB (2018). Photovoltaic plants. Technical application paper No. 10. www04. abb.com/global/seitp/seitp202.nsf/c71c66c1f02e6575c125711f004660e6/ d54672ac6e97a439c12577ce003d8d84/$file/vol.10.pdf (accessed 14 April 2018).

Adinoyi M. J. and Said S. A. M. (2013). Effect of dust accumulation on the power outputs of solar photovoltaic modules. *Renewable Energy*, **60**, 633–636, Elsevier.

Akhil A. A., Huff G., Currier A. B., Kaun B. C., Rastler D. M., Chen S. B., Cotter A. L., Bradshaw D. T. and Gauntlett W. D. (2013). DOE/EPRI 2013 Electricity Storage Handbook in Collaboration with NRECA. Sandia National Laboratories Albuquerque, New Mexico 87185 and Livermore, California 94550. www.sandia.gov/ess/publications/SAND2013–5131. pdf (accessed 14 April 2018).

Alkaisi A., Mossad R. and Sharifian-Barforoush A. (2017). A review of the water desalination systems integrated with renewable energy. *Energy Procedia*, **110**, 268–274, Elsevier.

Al-Karaghouli A. A. and Kazmerski L. L. (2011). Renewable Energy Opportunities in Water Desalination. Chapter 8 in Schorr (2011).

AMTA (2018). Membrane technology fact sheets. www.amtaorg.com/ publications-communications/membrane-technology-fact-sheets-summary (accessed 14 April 2018).

Asano T., Burton F. L., Leverenz H. L., Tsuchihashi R. and Tchobanoglous G. (2007). Water Reuse – Issues, Technologies, and Applications. Metcalf & Eddy/WECOM, McGraw-Hill, New York.

AWEA (2017). Wind Energy Reduces Greenhouse Gas Emissions. American Wind Energy Association. www.awea.org/reducing-greenhouse-gas-emissions (accessed 5 March 2018).

Balmér P. (2010). Benchmarking and Key Parameters in Wastewater Treatment Systems (in Swedish), Swedish Water, Report no 10, 2010. http://www.svensktvatten.se/rapporter/ (accessed 15 March 2018).

Bazargan, A. (ed.) (2018). A Multidisciplinary Introduction to Desalination, IWA Publishing, London.

Beckman K. (2016). Interview Adnan Amin, head of IRENA: "Everything we see is pointing to transformational change. Energy Post, http://energypost.eu/interview-adnan-amin-chief-irena-climate-negotiators-still-much-learn-energy-transformation (accessed 20 March 2018).

BNEF and Lighting Global (2016). Off-Grid Solar Market Trends Report, Commissioned by World Bank Group, Washington DC. https://www.lightingglobal.org/resource/off-grid-solar-market-trends-report-2016/ (accessed 15 February 2018).

Bundschuh J. and Hoinkis J. (eds) (2012). Renewable Energy Applications for Freshwater Production. IWA Publishing, London.

Burn S. and Gray S. (2014). Efficient Desalination by Reverse Osmosis: A Best Practice Guide to RO. IWA Publishing, London.

Campana P. E., Leduc S., Kim M., Olsson A., Zhang J., Liu J., Kraxner F., McCallum I., Li H. and Yan J. (2016). Suitable and optimal locations for implementing photovoltaic water pumping systems for grassland irrigation in China. *Applied Energy*, **185**, 1879–1889, Elsevier.

Campana P. E., Li H. and Yan J. (2015). Techno-economic feasibility of the irrigation system for the grassland and farmland conservation in China: Photovoltaic vs. wind power water pumping. *Energy Conversion and Management*, **103**, 311–320, Elsevier.

Casey A. (2013). "Reforming Energy Subsidies Could Curb India's Water Stress", http://www.worldwatch.org/reforming-energy-subsidies-could-curb-india%E2%80%99s-water-stress-0 (accessed 5 February 2018).

CEA (2016). Growth of Electricity Sector in India from 1947–2013. Central Electricity Authority, New Delhi, India, 2013. www.indiaenvironmentportal.org.in/content/380722/growth-of-electricity-sector-in-india-from-1947-2013/ (accessed 15 February 2018).

CEEW (2017). Council on Energy, Environment and Water. Greening India's Workforce: Gearing up for expansion of solar and wind power in India.

www.nrdc.org/sites/default/files/greening-india-workforce.pdf (accessed 15 February 2018).

Chandel S. S., Nagaraju Naik M. and Chandel R. (2017). Review of performance studies of direct coupled photovoltaic water pumping systems and case study. *Renewable and Sustainable Energy Reviews*, **76**, 163–175, Elsevier.

Cipollina A., Tzen E., Subiela V., Papapetrou M., Koschikowski J., Schwantes R., Wieghaus M. and Zaragoza G. (2015). Renewable energy desalination: Performance analysis and operating data of existing RES desalination plants. *Desalination and Water Treatment*, **55**, 3120–3140.

Climatescope (2017). 2018 off-grid and mini-grid market outlook. http://global-climatescope.org/en/ (accessed 13 March 2018).

Cust J., Singh A. and Neuhoff K. (2007). Rural Electrification in India: Economic and Institutional Aspects of Renewables. Cambridge Working Papers in Economics. University of Cambridge.

Daniels F. (1977). Direct Use of the Sun's Energy. 6th edn, Yale University Press, New Haven/London.

Delucchi M. A. and Jacobson M. Z. (2011). Providing all global energy with wind, water, and solar power, part II: Reliability, system and transmission costs, and policies. *Energy Policy*, **39**(3), 1170–1190, Elsevier.

DNV GL (2017). The energy transition outlook 2017. www.dnvgl.com/technology-innovation/sri/climate-action/research-projects/energy-transition-outlook.html (accessed 14 January 2018).

Drioli E., Criscuoli A. and Macedonio F. (2011). Membrane Based Desalination: An Integrated Approach. IWA Publishing, London.

EPA (2011). Environmental Protection Agency. Water treatment manual: Disinfection. www.epa.ie/pubs/advice/drinkingwater/Disinfection2_web.pdf (accessed 15 February 2018).

Espino T., Peñate B., Piernavieja G., Herold D. and Neskakis A. (2003). Optimised desalination of seawater by a PV powered reverse osmosis plant for a decentralised coastal water supply. *Desalination*, **156**(1–3), 349–350.

Fraunhofer (2015). Fraunhofer Institute for Solar Energy Systems, NREL. Current status of concentrator photovoltaic (CPV) technology, http://www.nrel.gov/docs/fy16osti/65130.pdf (accessed 1 February 2018).

Fraunhofer (2016). Fraunhofer Institute for Solar Energy Systems, Photovoltaics Report, June 6, https://www.ise.fraunhofer.de/de/downloads/pdf-files/aktuelles/photovoltaics-report-inenglischer-sprache.pdf (accessed 1 February 2018).

Friedler E., Butler D. and Alfiya Y. (2013). Wastewater composition. Chapter 17 in Larsen *et al.* (2013).

Gillblad T. (2018). Power requirement for a decentralised activated sludge plant. Personal communication.

GOGLA (2017). Global Off-Grid Lighting Association. Providing Energy Access through Off-Grid Solar: Guidance for Governments. The Netherlands, www.se4all.org/content/new-report-guides-governments-designing-off-grid-solar-policies (accessed 22 March 2018).

Goswani D. Y. (2015). Principles of Solar Engineering. 3rd edn, CRC Press, Taylor & Francis Group, USA.

Grady C. P. L., Daigger G. T., Love N. G. and Filipe, C. D. M. (2011). Biological Wastewater Treatment. 3rd edn, IWA Publishing and CRC Press, London.

Greacen C., Engel R. and Quetchenbach T. (2013). A Guidebook on Grid Interconnection and Islanded Operation of Mini-Grid Power Systems Up to 200 kW. Lawrence Berkeley National Laboratory and Schatz Energy Research Center, LBNL-6224E, http://etapublications.lbl.gov/sites/default/files/a_guidebook_for_minigrids-serc_lbnl_march_2013.pdf (accessed 14 April 2018).

Green J. (2018). Nuclear power in crisis: We are entering the Era of Nuclear Decommissioning. Energy Post, webpage for European Energy Affairs, http://energypost.eu/nuclear-power-in-crisis-welcome-to-the-era-of-nuclear-decommissioning (accessed 25 March 2018).

Gude V. G., Nirmalakhandan N. and Deng S. (2010). Renewable and sustainable approaches for desalination. *Renewable and Sustainable Energy Reviews*, **14**, 2641–2654, Elsevier.

Hartvigsson E. and Ahlgren E. O. (2018). Comparison of load profiles in a mini-grid: Assessment of performance metrics using measured and interview-based data. *Energy for Sustainable Development*, **43**, 186–195. doi:10.1016/j.esd.2018.01.009.

Hoffman A. R. (2016). The U.S. Government and Renewable Energy – A Winding Road. Pan Stanford Publishing Pte. Ltd., Singapore.

Hoffman A. R. (2017a). Important large market: Solar energy and clean water. Chapter 2.6 in Varadi (2017).

Hoffman A. R. (2017b). Subsidies and solar energy. Chapter 3.2 in Varadi (2017).

Huld T., Jäger-Waldau A. and Szabó S. (2014). European Commission Joint Research Centre. Mapping the Cost of Electricity from Grid-Connected and Off-Grid PV Systems in Africa. 1st Africa Photovoltaic Solar Energy Conference, Durban, South Africa. www.researchgate.net/publication/263221188_Mapping_the_Cost_of_Electricity_from_Grid-connected_and_Off-grid_PV_Systems_in_Africa (accessed 15 April 2018).

IAEA (2017). International Atomic Energy Agency, International Status and Prospects for Nuclear Power, www.iaea.org/About/Policy/GC/GC61/GC61InfDocuments/English/gc61inf-8_en.pdf (accessed 15 February 2018).

IEA (2011). Energy for All – Financing Access for the Poor. World Energy Outlook, Special Issue. International Energy Agency, Paris. www.iea.org/media/weowebsite/energydevelopment/presentation_oslo_oct11.pdf (accessed 15 February 2018).

IEA (2013). Rural Electrification with PV Hybrid Systems. Report IEA-PVPS T9-13:2013, International Energy Agency XE "International Energy Agency", Paris. www.iea-pvps.org/fileadmin/dam/public/report/national/Rural_Electrification_with_PV_Hybrid_systems_-_T9_-_11072013_-_Updated_Feb2014.pdf (accessed 10 April 2018).

IEA (2014). World Energy Outlook 2014, International Energy Agency, Paris.

IEA (2016a). World Energy Outlook 2016, International Energy Agency, Paris, 2016.

IEA (2016b). Next Generation Wind and Solar Power: From Cost to Value. International Energy Agency, Paris. https://www.iea.org/publications/freepublications/publication/NextGenerationWindandSolarPower.pdf (accessed 10 April 2018).

IEA (2017a). Key World Energy Statistics. International Energy Agency, Paris. www.iea.org/publications/freepublications/publication/KeyWorld2017.pdf (accessed 15 February 2018).

IEA (2017b). Renewables 2017. Analysis and forecasts to 2022. International Energy Agency, Paris. www.iea.org/renewables/ (accessed 11 February 2018).

IEA (2017c). Energy Access Outlook 2017. From Poverty to Prosperity. World Energy Outlook Special Report. http://www.iea.org (accessed 11 February 2018).

IEA (2017d). Africa Energy Outlook. A focus on energy prospects in sub-Saharan Africa. www.iea.org (accessed 1 May 2018).

IEA-ETSAP and IRENA (2013). Water Desalination Using Renewable Energy, Technology Brief I12, https://iea-etsap.org/E-TechDS/PDF/I12IR_Desalin_MI_Jan2013_final_GSOK.pdf (accessed 22 January 2018).

IEA India (2015). India Energy Outlook, World Energy Outlook Special Report. http://www.worldenergyoutlook.org/media/weowebsite/2015/India EnergyOutlook_WEO2015.pdf (accessed 12 February 2018).

IEC (2011). Electrical energy storage, a White Paper. www.iec.ch/whitepaper/pdf/iecWP-energystorage-LR-en.pdf (accessed 15 April 2018).

Inversin A. R. (2000). Mini-Grid Design Manual. National Rural Electric Cooperative Association (NRECA), Arlington, Virginia, U.S., www.reseau-cicle.org/wp-content/uploads/riaed/pdf/Mini-Grid_Design_Manual-2.pdf (accessed 14 April 2018).

IPCC (2011). IPCC Special Report on Renewable Energy Sources and Climate Change Mitigation. Prepared by Working Group III of the Intergovernmental Panel on Climate Change [Edenhofer O., Pichs-Madruga R., Sokona Y., Seyboth K., lead authors]. Cambridge University Press, Cambridge, UK and New York, NY, USA, 1075 pp (Chapter 7 & 9). www.ipcc.ch/pdf/special-reports/srren/SRREN_FD_SPM_final.pdf (accessed 23 February 2018).

IRENA (2012). International Off-grid Renewable Energy Conference 2012: Key Findings and Recommendations, Accra, Ghana, 1–2 November 2012. Available at www.irena.org/DocumentDownloads/Publications/IOREC_Key20Findings20and%20Recommendations.pdf (accessed 20 February 2018).

IRENA (2013). Africa's renewable future. The path to sustainable growth. www.irena.org/documentdownloads/publications/africa_renewable_future.pdf (accessed 30 April 2018).

IRENA (2015a). Accelerating Off-grid Renewable Energy Deployment: Key Findings and Recommendations from IOREC 2014. IRENA, Abu Dhabi. www.irena.org/DocumentDownloads/Publications/IRENA_2nd_IOREC_2015.pdf (accessed 15 February 2018).

IRENA (2015b). Renewable Energy in the Water, Energy and Food Nexus. IRENA, Abu Dhabi. www.irena.org/documentdownloads/publications/irena_water_energy_food_nexus_2015.pdf (accessed 20 February 2018).

IRENA (2015c). Quality Infrastructure for Renewable Energy Technologies: Small Wind Turbines, www.irena.org/publications/2015/Dec/Quality-Infrastructure-for-Renewable-Energy-Technologies-Small-Wind-Turbines (accessed 20 March 2018).

IRENA (2016a). Renewable Capacity Statistics 2016. Abu Dhabi. www.irena.org/DocumentDownloads/Publications/IRENA_RE_Capacity_Statistics_2016.pdf (accessed 19 February 2018).

IRENA (2016b). Letting in the Light. How Solar Photovoltaics will Revolutionize the Electricity System. International Renewable Energy Agency, Abu Dhabi. www.irena.org/DocumentDownloads/Publications/IRENA_Letting_in_the_Light_2016.pdf (accessed 20 March 2018).

IRENA (2016c). REmap: Roadmap for A Renewable Energy Future: 2016 Edition, www.irena.org/publications/2016/Mar/REmap-Roadmap-for-A-Renewable-Energy-Future-2016-Edition (accessed 20 March 2018).

IRENA (2016d). Solar PV in Africa: Costs and Markets, www.irena.org/publications/2016/Sep/Solar-PV-in-Africa-Costs-and-Markets (accessed 15 March 2018).

IRENA (2016e). Solar Pumping for Irrigation: Improving Livelihoods and Sustainability. IRENA, Abu Dhabi. www.irena.org/publications/2016/Jun/Solar-Pumping-for-Irrigation-Improving-livelihoods-and-sustainability (accessed 15 March 2018).

IRENA (2017a). REthinking Energy 2017: Accelerating the Global Energy Transformation. International Renewable Energy Agency, Abu Dhabi. ISBN 978-92-95111-05-9 (print) | ISBN 978-92-95111-06-6 (PDF). www.irena.org/DocumentDownloads/Publications/IRENA_REthinking_Energy_2017.pdf (accessed 10 February 2018).

IRENA (2017b). Electricity storage and renewables: Costs and markets to 2030, www.irena.org/publications/2017/Oct/Electricity-storage-and-renewables-costs-and-markets (accessed 15 March 2018).

IRENA (2017c). Renewable energy and jobs. Annual review 2017 www.irena.org/DocumentDownloads/Publications/IRENA_RE_Jobs_Annual_Review_2017.pdf (accessed 15 April 2018).

IRENA (2018). Renewable Power Generation Costs in 2017, www.irena.org/publications/2018/Jan/Renewable-power-generation-costs-in-2017 (accessed 20 March 2018).

Jones L. E. (ed.) (2017). Renewable Energy Integration: Practical Management of Variability, Uncertainty and Flexibility in Power Grids. 2nd edn, Academic Press, published by Elsevier Inc., Amsterdam, The Netherlands.

Jones L. E. (2018). Utilities and Business: How Scaling Up Energy Access Delivers 'Wins' for Both. BRINK News www.brinknews.com/utilities-and-business-how-scaling-up-energy-access-delivers-wins-for-both/ (accessed 15 February 2018).

Jones L. E., Akycampong E. K., Kutarna C. and Jasieniak J. J. (2018). Energy for Education Is Vital for Developing Economies. BRINK News. http://www.brinknews.com/energy-for-education-is-vital-for-developing-economies/ (accessed 15 February 2018).

Jones L. E. and Olsson G. (2017). Solar PV and wind energy providing water. *Global Challenges*, special issue on water and energy, **1**(5), 8 pages. Open access, https://onlinelibrary.wiley.com/doi/pdf/10.1002/gch2.201600022 (accessed 1 March 2018).

Kalogirou S. A. (2004). Solar thermal collectors and applications. *Progress in Energy and Combustion Science*, **30**(3), 231–295, Elsevier.

Kalogirou S. A. (2005). Seawater desalination using renewable energy sources. *Progress in Energy and Combustion Science*, **31**, 242–281, Elsevier.

Kaur T. and Segal R. (2017). Designing rural electrification solutions considering hybrid energy systems for Papua New Guinea. *Energy Procedia*, **110**, 1–7. www.sciencedirect.com.

Kirubi C., Jacobson A., Kammen D. M. and Mills A. (2009). Community-based electric micro-grids can contribute to rural development: Evidence from Kenya. *World Development*, **37**(7), 1208–1221.

Kittner N., Lill F. and Kammen D. M. (2017). Energy storage deployment and innovation for the clean energy transition. *Nature Energy*, **2**, 17125. https://rael.berkeley.edu/project/innovation-in-energy-storage/ (accessed 10 April 2018).

Kumar A., Schei T., Ahenkorah A., Caceres Rodriguez R., Devernay J.-M., Freitas M., Hall D., Killingtveit Å. and Liu Z. (2011). Hydropower. In: IPCC Special Report on Renewable Energy Sources and Climate Change Mitigation, Edenhofer O., Pichs-Madruga R., Sokona Y., Seyboth K., Matschoss P., Kadner S., Zwickel T., Eickemeier P., Hansen G., Schlömer S. and von Stechow C. (eds), Cambridge University Press, Cambridge, United Kingdom and New York, NY, USA, pp. 437–496.

Lapsed Physicist (2018). Allan R. Hoffman blog: Thoughts of a Lapsed Physicist: Perspectives on energy and water technologies and policy, www.lapsedphysicist.org.

Larminie J. and Dicks A. (2003). Fuel Cell Systems Explained. 2nd edn, John Wiley & Sons Ltd., Chichester, West Sussex, England. http://eng.harran.edu.tr/moodle/moodledata/134/Ders_Notlari/fuel_pilleri.pdf (accessed 6 April 2018).

Larsen T., Udert K. M. and Lienert J. (eds) (2013). Source Separation and Decentralization for Wastewater Management. IWA Publishing, London.

Lavelle M. (2015). Can sun and wind make more salt water drinkable? *National Geographic*, February 2. http://news.nationalgeographic.com/news/energy/2015/02/150202-energy-news-renewable-salt-water-drought/ (accessed 5 February 2018).

Lazard (2016). Lazard's levelized cost of storage, version 2.0. www.lazard.com/media/438042/lazard-levelized-cost-of-storage-v20.pdf (accessed 1 March 2018).

Lazard (2017). Lazard's levelized cost of energy analysis, version 11, Nov. 2017. www.lazard.com/perspective/levelized-cost-of-energy-2017/ (accessed 5 March 2018).

Leslie G. and Bradford-Hartke Z. (2013). Membrane processes. Chapter 25 in Larsen *et al.* (2013).

Liehr S., Kramm J., Jokisch A. and Müller K. (eds) (2018). Integrated Water Resources Management in Water-Scarce Regions: Water Harvesting, Groundwater Desalination and Water Reuse in Namibia. IWA Publishing, London.

Lienhard J. H., Thiel G. P., Warsinger D. M. and Banchik L. D. (eds) (2016). Low Carbon Desalination: Status and Research, Development, and Demonstration Needs. Report of a Workshop Conducted at the Massachusetts Institute of Technology in Association with the Global Clean Water Desalination Alliance. MIT Abdul Latif Jameel World Water and Food Security Lab, Cambridge, Massachusetts, November 2016, pp. 138. http://web.mit.edu/lowcdesal/ (accessed 25 February 2018).

Lighting Africa (2013). Lighting Africa: Market Trends Report 2012, Lighting Africa & The World Bank Group, www.lightingafrica.org/wp-content/uploads/2016/07/5_Market-Brief-Report-ElectronicREV-1.pdf (accessed 15 February 2018).

Lingsten A., Lundqvist M., Hellström D. and Balmér P. (2011). Energy use in Swedish wastewater treatment plants 2008 (in Swedish). Swedish Water Report no 4, 2011. www.svensktvatten.se/rapporter/ (accessed 15 March 2018).

Mahmoudi H., Ghaffour N., Goosen M. F. A. and Bundschuh J. (2017). Renewable Energy Technologies for Water Desalination. CRC Press, Taylor & Francis Group, USA.

Manwell J. F., McGowan J. G. and Rogers A. L. (2011). Wind Energy Explained: Theory, Design and Application. 2nd edn, John Wiley & Sons, Ltd, UK.

Masdar (2018). Renewable energy water desalination programme. The New Frontier of Sustainable Water Desalination, Report http://masdar.ae/assets/downloads/content/3588/desalination_report-2018.pdf (accessed 20 March 2018).

Mathews J. A. (2016). Competing principles driving energy futures: Fossil fuel decarbonization vs. manufacturing learning curves. *Futures*, **84**(A), 1–11.

Mathews J. A. and Huang X. (2018). China's green energy revolution has saved the country from catastrophic dependence on fossil fuel imports, *Energy Post*, webpage for European Energy Affairs, http://energypost.eu/chinas-green-energy-revolution-has-saved-the-country-from-catastrophic-dependence-on-fossil-fuelimports/ (accessed 28 March 2018).

Mathews J. A. and Tan H. (2014). Economics: Manufacture renewables to build energy security. *Nature*, **513**, 166–168. 11 Sep. 2014 Comment.

Metcalf and Eddy (2013). Wastewater Engineering – Treatment and Reuse. 5th edn, McGraw-Hill Education, USA.

NREL (2012a). Life Cycle Greenhouse Gas Emissions from Solar Photovoltaics. National Renewable Energy Laboratory, USA. www.nrel.gov/docs/fy13osti/56487.pdf (accessed 23 February 2018).

NREL (2012b). Renewable Electricity Futures Study. National Renewable Energy Laboratory, USA. www.nrel.gov/docs/fy13osti/52409-ES.pdf (accessed 23 February 2018).

Olsson G. (2015). Water and Energy: Threats and Opportunities. 2nd edn, IWA Publishing, London.

Oram B. (2014). UV Disinfection Drinking Water. Water Research Center, Dallas, PA, USA. https://www.water-research.net/index.php/water-treatment/water-disinfection/uv-disinfection (accessed 23 February 2018).

Palit D. and Chaurey A. (2011). Off-grid rural electrification experiences from South Asia: Status and best practices. *Energy for Sustainable Development*, **15**(3), 266–276. doi:10.1016/j.esd.2011.07.004.

Parida B., Iniyan S. and Goic R. (2011). A review of solar photovoltaic technologies. *Renewable and Sustainable Energy Reviews*, **15**, 1625–1636, Elsevier.

Piani G. (2017). Python Solar Energy Calculation Primer. Code and manuscript, download at https://pypi.python.org/pypi/solprimer/1.0.1.

Prinsloo G., Dobson R. and Brent A. (2016). Scoping exercise to determine load profile archetype reference shapes for solar co-generation models in isolated off-grid rural African villages. *Journal of Energy in Southern Africa*, **27**(3), 11–27. www.scielo.org.za/pdf/jesa/v27n3/02.pdf (accessed 1 February 2018).

Rahman A. and Bhatt B. P. (2014). Design approach for solar photovoltaic groundwater pumping system for Eastern India. *Current World Environment*, **9**(2), 426–429, www.cwejournal.org/pdf/vol9no2/vol9_no2_426-429.pdf (latest approach 15 February 2018).

REN21 (2017a). Renewable Energy Network for the 21st Century. Renewables 2017, Global Status Report. http://www.ren21.net/status-of-renewables/global-status-report/ (accessed 20 February 2018).

REN21 (2017b). Renewables Global Futures Report: Great debates Towards 100% Renewable Energy.REN21 Secretariat, Paris. ISBN 978-3-9818107-4-5.

Rever B. (2017). Utility-scale solar. Chapter 2.5 in Varadi (2017).

Riva F., Ahlborg H., Hartvigsson E., Pachauri S. and Colombo E. (2018). Electricity access and rural development: Review of complex socio-economic dynamics and casual diagrams for more appropriate energy modelling. *Energy for Sustainable Development*, **43**, 203–223, www.sciencedirect.com.

Roul A. (2007). India's Solar Power – Greening India's Future Energy Demand. ECO World, https://web.archive.org/web/20080619165847/http://www.ecoworld.com/ (accessed 15 February 2018).

Schnitzer D., Lounsbury D. S., Carvallo J. P., Deshmukh R., Apt J. and Kammen D. M. (2014). Microgrids for Rural Electrification: A Critical Review of Best Practices Based on Seven Case Studies. Published by the UN Foundation, Washington D.C. http://energyaccess.org/wp-content/uploads/2015/07/MicrogridsReportFINAL_high.pdf (accessed 15 April 2018).

Schorr M. (ed.) (2011). Desalination, Trends and Technologies. IntechOpen Limited, London.

Shatat M., Worall M. and Riffat S. (2013). Opportunities for solar water desalination worldwide: Review. *Sustainable Cities and Society*, **9**, 67–80. Elsevier.

Shukla O., Agravat S. M., Jani B. B., Srivastava N. and Singh, G. (2016). Canal Top Solar PV Plant in Gujarat. A Unique Nexus of Energy, Land, and Water Akshay Urja, India, August 2016, 20–23, http://mnre.gov.in/file-manager/akshay-urja/july-august-2016/EN/20-23.pdf (accessed 5 February 2018).

SNV (2015). Solar PV standardised training manual. www.snv.org/public/cms/sites/default/files/explore/download/standardised-training-manual.pdf (accessed 15 April 2018).

Sontake V. C. and Kalamkar V. R. (2016). Solar photovoltaic water pumping system – a comprehensive review. *Renewable and Sustainable Energy Reviews*, **59**, 1038–1067, Elsevier.

Stenström T. A. (2013). Hygiene, a major challenge for source separation and decentralization. Chapter 11 in Larsen *et al.* (2013).

Svensson J. (2018). Wind power, personal communication. Dr. J. Svensson, Lund University.

Swift A., Rainwater K., Chapman J., Noll D., Jackson A., Ewing B., Song L., Ganesan G., Marshall R., Doon V. and Nash P. (2009). Wind Power and Water Desalination Technology Integration, Technical Report, Texas Tech University, Lubbock, Texas & U.S. Department of the Interior. Available at https://www.usbr.gov/research/dwpr/reportpdfs/report146.pdf (accessed 8 February 2018).

Tilley E. (2013). Conceptualizing sanitation systems to account for new complexities in processing and management. Chapter 16 in Larsen *et al.* (2013).

Tweed K. (2014). India Plans to Install 26 million Solar-powered Water Pumps, http://spectrum.ieee.org/energywise/green-tech/solar/india-plans-for-26-million-solar-water-pumps (accessed 5 February 2018).

Uhrig M., Koenig S., Suriyah M. R. and Leibfried, T. (2016). Lithium-based vs. vanadium redox flow batteries – a comparison for home storage systems. *Energy Procedia*, **99**, 35–43, Elsevier.

UN Water (2014). *Water for Life 2005–2015*, United Nations International Decade for Action. www.un.org/waterforlifedecade/africa.shtml (accessed 5 February 2018).

UN WWDR (2014). Water and Energy, The United Nations World Water Development Report 2014 (2 volumes). UN World Water Assessment Programme. Unesco, Paris. http://unesdoc.unesco.org/images/0022/002257/225741E.pdf (accessed 18 February 2018).

USAID (2014). Hybrid mini-grids for rural electrification: Lessons learned. USAID, prepared by the Alliance for Rural Electrification.

https://ruralelec.org/sites/default/files/hybrid_mini-grids_for_rural_ electrification_2014.pdf (accessed 13 April 2018).

Varadi P. F. (2017). Sun Towards High Noon: Solar power transforming our energy future. Pan Stanford Series on Renewable Energy, vol 8., Pan Stanford Publishing, ISBN 978-981-4774-17-8.

Varadi P. F., Wouters F. and Hoffman A. R. (2018). The sun is rising in Africa and the Middle East. On the road to a solar energy future. Pan Stanford Series on Renewable Energy, vol 9., Pan Stanford Publishing, ISBN 978-981-4774-89-5.

Voutchkov N. (2012). Desalination Engineering: Planning and Design. WEF Press & McGraw Hill, Alexandria, VA, USA.

Voutchkov N. (2016). Desalination – Past, Present and Future. The Source, International Water Association. London, UK.

WEC (2016). World Energy Council. World Energy Resources 2016, Chapter 8. Khetarpal, D (lead author). Available at https://www.worldenergy. org/wp-content/uploads/2016/10/World-Energy-Resources-Full-report-2016.10.03.pdf (accessed 5 February 2018).

Weiner D., Fisher D., Katz B, Moses E. J. and Meron, G. (2001). Operation experience of a solar- and wind-powered desalination demonstration plant. *Desalination*, **137**. 7–13, Elsevier.

Wesoff E. and Lacey S. (2017). Solar costs are hitting jaw-dropping lows in every region of the world, *GTM Research*, www.greentechmedia.com/ articles/read/Solar-Costs-Are-Hitting-Jaw-Dropping-Lows-In-Every-Region-of-the-World#gs.8UA3X0Y (accessed 20 March 2018).

White S. (2014). Solar Photovoltaic Basics, Routledge.

White S. (2016). Solar PV Engineering and Installation. Routledge, London, UK.

WHO (2006). Guidelines for the safe use of wastewater, excreta and greywater. Vol. 4: Excreta and greywater use in agriculture. World Health Organization, Geneva, Switzerland. www.who.int/water_sanitation_ health/publications/gsuweg4/en/ (accessed 15 March 2018).

WorldBank (2014). Electric Power consumption. World Bank Data. https:// data.worldbank.org/indicator (accessed 10 February 2018).

World Bank (2017). State of Electricity Access Report 2017. https://open knowledge.worldbank.org/handle/10986/26646 (accessed 15 March 2018).

WWEA (2016). Small Wind World Report 2016, World Wind Energy Association, Bonn, Germany. www.wwindea.org/ (accessed 20 February 2018).

Yadoo A. and Cruickshank H. (2010). The value of cooperatives in rural electrification. *Energy Policy*, **38**(6), 2941–2947. doi:https://doi.org/ 10.1016/j.enpol.2010.01.031.

Zabbey N. and Olsson G. (2017). Conflicts – oil exploration and water. *Global Challenges*, **1**(5), 10 pages. Open access, Special issue on Water and energy. http://onlinelibrary.wiley.com/doi/10.1002/gch2.201600015/full (accessed 20 February 2018).

Zaragoza G. (2018). Private communication. CIEMAT Plataforma Solar de Almeria, Almeria, Spain.

Index

Recommended titles by Gustaf Olsson

Smart Water Utilities: Complexity Made Simple
Pernille Ingildsen & Gustaf Olsson
ISBN: 9781780407579
May 2016
Standard price: £79 / US$119 / €99
Member price: £59 / US$89 / €74

Water and Energy: Threats and Opportunities - Second Edition
Gustaf Olsson
ISBN: 9781780406930
June 2016
Standard price: £103 / US$155 / €129
Member price: £77 / US$116 / €96

This book provides a framework for Smart Water Utilities based on an M-A-D (Measurement-Analysis-Decision).This enables the organisation and implementation of "Smart" in a water utility by providing an overview of supporting technologies and methods.

This new edition of *Water and Energy: Threats and Opportunities* is timely and continues to highlight the inextricable link between water and energy, providing an up-to-date overview of the subject with helpful detailed summaries of the technical literature.

With contributions by Gustaf Olsson

Aeration, Mixing, and Energy: Bubbles & Sparks
Diego Rosso
ISBN: 9781780407838
December 2018
Standard price: £89 / US$134 / €111
Member price: £67 / US$101 / €84

Advances in Wastewater Treatment
Giorgio Mannina, George Ekama, Hallvard Ødegaard & Gustaf Olsson
ISBN: 9781780409702
October 2018
Standard price: £140 / US$210 / €175
Member price: £105 / US$158 / €131

This valuable book compiles existing knowledge on aeration, mixing, and their energy implications for water reclamation and wastewater treatment. A valuable complement to any book on water reclamation and wastewater treatment.

A compendium of key topics, future technologies and perspectives in wastewater treatment and modelling. It covers the fundamentals and innovative treatment processes and focuses on the key role of modelling in design, operation and control.

Activated Sludge - 100 Years and Counting
David Jenkins & Jiri Wanner
ISBN: 9781780404936
May 2014
Standard price: £146 / US$219 / €183
Member price: £110 / US$165 / €138

Biological Wastewater Treatment
Mogens Henze, Mark C.M. van Loosdrecht, G.A. Ekama & Damir Brdjanovic
ISBN: 9781843391883
September 2008
Standard price: £123 / US$185 / €154
Member price: £92 / US$138 / €115

This celebrated book covers the current status of all aspects of the activated sludge process and looks forward to its further development in the future.

IWA Publishing's best-selling title comprehensively covers advances in wastewater treatment. Also available in Spanish.

IWA Publishing's authorised EU representative for General Product Safety Regulations is Diane D'Arras, 15 rue Duret, 75116 Paris, France, e-mail: safety@iwap.co.uk.

Printed and bound by CPI Group (UK) Ltd, Croydon, CR0 4YY

12/05/2026

02108891-0001